'*Governing the Environment* presents us with diverse and innovative scholarship on how early modern thinkers interpreted the complex relationships between people and their dynamic environments. Although focused on the past, this well-crafted volume provides fresh perspectives on current interrogations into what constitutes "nature" in light of the long history of politicized climate knowledge, the variable effects of human agency, and the challenges of environmental governance projects.'

Mary Floyd-Wilson, *University of North Carolina, USA*

'With learning lightly worn, these insightful essays illuminate the multiple, and ever-evolving, understandings of climate and the environment circulating in Western Europe and North America in the early modern centuries. They convincingly show how deeply environmental ideas, and management practices, were embedded in prevailing political and social orders – then as now.'

John McNeill, *Georgetown University, USA*

Governing the Environment in the Early Modern World

Throughout the early modern period, scientific debate and governmental action became increasingly preoccupied with the environment, generating discussion across Europe and the wider world as to how to improve land and climate for human benefit. This discourse eventually promoted the reconsideration of long-held beliefs about the role of climate in upholding the social order, driving economies and affecting public health.

Governing the Environment in the Early Modern World explores the relationship between cultural perceptions of the environment and practical attempts at environmental regulation and change between 1500 and 1800. Taking a cultural and intellectual approach to early modern environmental governance, this edited collection combines an interpretative perspective with new insights into a period largely unfamiliar to environmental historians. Using a rich and multifaceted narrative, this book offers an understanding as to how efforts to enhance productive aspects of the environment were both led by and contributed to new conceptualisations of the role of 'nature' in human society.

This book offers a cultural and intellectual approach to early modern environmental history and will be of special interest to environmental, cultural and intellectual historians, as well as anyone with an interest in the culture and politics of environmental governance.

Sara Miglietti is an Assistant Professor of French Studies at Johns Hopkins University, Baltimore, USA.

John Morgan is an environmental and social historian, and a Lecturer in Early Modern History at the University of Manchester, UK.

Routledge Environmental Humanities
Series editors: Iain McCalman and Libby Robin

The *Routledge Environmental Humanities* series is an original and inspiring venture recognising that today's world agricultural and water crises, ocean pollution and resource depletion, global warming from greenhouse gases, urban sprawl, overpopulation, food insecurity and environmental justice are all *crises of culture.*

The reality of understanding and finding adaptive solutions to our present and future environmental challenges has shifted the epicenter of environmental studies away from an exclusively scientific and technological framework to one that depends on the human-focused disciplines and ideas of the humanities and allied social sciences.

We thus welcome book proposals from all humanities and social sciences disciplines for an inclusive and interdisciplinary series. We favour manuscripts aimed at an international readership and written in a lively and accessible style. The readership comprises scholars and students from the humanities and social sciences and thoughtful readers concerned about the human dimensions of environmental change.

Governing the Environment in the Early Modern World

Theory and Practice

Edited by Sara Miglietti and John Morgan

Routledge
Taylor & Francis Group
LONDON AND NEW YORK

earthscan
from Routledge

First published 2017
by Routledge

2 Park Square, Milton Park, Abingdon, Oxfordshire OX14 4RN
52 Vanderbilt Avenue, New York, NY 10017

Routledge is an imprint of the Taylor & Francis Group, an informa business

First issued in paperback 2018

British Library Cataloguing-in-Publication Data
A catalogue record for this book is available from the British Library

Library of Congress Cataloging-in-Publication Data
Names: Miglietti, Sarah, editor. | Morgan, John (Johm Emrys), editor.
Title: Governing the environment in the early modern world : theory and practice / edited by Sarah Miglietti and John Morgan.
Description: Abingdon, Oxon ; New York, NY : Routledge, 2017.
Identifiers: LCCN 2016044318 | ISBN 9781138674776 (hbk) | ISBN 9781315561097 (ebk)
Subjects: LCSH: Human ecology—History. | Nature—Effect of human beings on—History. | Global environmental change—History. | Environmental policy—History.
Classification: LCC GF13 .G68 2017 | DDC 304.209/03—dc23
LC record available at https://lccn.loc.gov/2016044318

ISBN: 978-1-138-67477-6 (hbk)
ISBN: 978-0-367-15232-1 (pbk)

Typeset in Bembo
by Apex CoVantage, LLC

Contents

Contributors

Anthony E. Carlson is an Assistant Professor of history at the School of Advanced Military Studies in Fort Leavenworth, Kansas. He has published several articles and chapters about American and transnational water, drainage and irrigation policy. He is currently completing a manuscript about American wetlands drainage policy and political development from the American Revolution to World War I.

William M. Cavert is an Assistant Professor of history at the University of St. Thomas, Minnesota. He is a historian of Britain during the early modern period, c. 1500–1800, with research interests in urban and environmental history. He is the author of *The Smoke of London: Energy and Environment in the Early Modern City*, published in 2016 by Cambridge University Press, as well as studies in *The Journal of British Studies*, *Urban History* and *The Global Environment*.

Ashley Dodsworth is a Teaching Associate in the Department of Politics and Sociology at the University of Bristol. She is interested in the intersection of environmental politics and the history of political thought, as seen in her PhD thesis, which examined past conceptions of environmental rights. She is currently researching how current environmentalists engage with Gerrard Winstanley and the relationship between the work of Mary Wollstonecraft and the eco-feminist tradition (Ashley.Dodsworth@bristol.ac.uk).

Rebecca Earle teaches history at the University of Warwick. She is the author of several books on Spanish American cultural and political history. Her current research explores the cultural significance of food and eating and the impact of New World foods on early modern European mentalities and political culture.

Mike Hulme is Professor of Climate and Culture and Head of the Department of Geography in the Faculty of Social Science & Public Policy at King's College London. His research explores the idea of climate change using historical, cultural and scientific analyses, seeking to illuminate the numerous ways in which climate change is deployed in public and political discourse. He is the author of more than 200 articles and numerous books, including *Why We Disagree About Climate Change* (Cambridge University Press, 2009).

Sara Miglietti is an Assistant Professor of French Studies at Johns Hopkins University. She specializes in the intellectual history of early modern France and Europe, with a focus on natural philosophy, medicine and political thought. Her publications include a commented edition of Jean Bodin's *Methodus ad facilem historiarum cognitionem* (Pisa: Edizioni della Normale, 2013) and the edited collection *Reading Publics in Renaissance Europe* (with Sarah E. Parker). She is currently completing a monograph on perceptions of environmental influence from the Renaissance to the Enlightenment (smiglie1@jhu.edu).

Raphaël Morera is a *chargé de recherche*, researcher, at the *Centre de recherches historiques* at the *École des hautes études en sciences sociales*. His research interests include the history of drainage and comparative histories of water management. His monograph, *L'assèchement des marais en France au XVIIe siècle*, was published by the University of Rennes Press in 2011.

John Morgan is an environmental and social historian and a lecturer in early modern history at the University of Manchester. He has published on the cultural histories of fires and flooding in early modern England and the Elizabethan religious settlement. He is currently researching the history of water management in England and the impact of coastal flooding on economic life in early modern Lincolnshire (John.Morgan@manchester.ac.uk).

Leona Skelton is an environmental historian of rivers and sanitation infrastructure and is Vice Chancellor's Research Fellow in the Humanities at Northumbria University in Newcastle. Her work focuses on the two-way interactions between people and the environment, the developments in environmental attitudes and regulations and how dramatic environmental change has shaped economic, cultural and social lives and livelihoods across northern England and Scotland between 1500 and the present day. She has published two monographs: *Sanitation in Urban Britain, 1560–1700* (London: Routledge, 2015) and *Tyne after Tyne: An Environmental History of a River's Battle for Protection, 1530–2015* (Cambridge: White Horse Press, 2017) (leona.skelton@northumbria.ac.uk).

Richard Spavin is currently a Postdoctoral Research Fellow at the Université de Montréal. His research interests focus on the culture and politics of early modern science in France. His forthcoming monograph, *Les climats du pouvoir: Bodin, Montesquieu, Rousseau* (Oxford Studies in the Enlightenment), analyses how political philosophers used 'climate theories' to conceptualise the moral imperatives of government. His current research project, funded by the Social Sciences and Humanities Research Council of Canada, explores the cultural history of mining and metals in early French modernity.

Anya Zilberstein is Associate Professor in the Department of History at Concordia University in Montréal and the author of *A Temperate Empire: Making Climate Change in Early America* (New York: Oxford University Press, 2016).

Foreword

The idea of anthropogenic climate change, that human activities are changing the patterns of weather worldwide, has become iconic for our age. It is an idea that not merely describes our present condition; it also *interprets* the human condition and through this interpretation shapes its future. This influence is achieved not least through the way in which the idea of climate change has prompted the conception that we are now living in the epoch of 'the Anthropocene' – the age of humans. The Anthropocene opens up a new repertoire of imaginative possibilities about the future.

But anthropogenic climate change is not just about the present and the future. It also is an idea that can change the way we think about, write about and understand the *past*. Like the future, the past is malleable. It is sensitive and responsive to how contemporary ideologies and idioms are used to select, synthesise and impart meaning to historical traces of human practices and non-human processes. History matters, environmental history especially. New historical narratives shaped by contemporary prejudices are needed to challenge an easy presumption that the future is merely about 'the future', that the arrow of time is linear. Paradoxically, it is often a historical perspective on contemporary discourses or problems, which better equips us – or at least differently equips us – to handle the future than do the predictive tools and instruments of scientific or political analysis.

As I have written elsewhere, the idea of climate change performs useful political and cultural work, here too in stimulating new histories of the past. Miglietti and Morgan's book *Governing the Environment in the Early Modern World* opens out a series of fresh perspectives on the contemporary politics of climate knowledge and the conundrums of climate governance. But it does so indirectly, by taking a backward look at how early modern Europeans and Americans grappled with the challenges of living in a physical world which was less than optimal and seemingly uncontrollable. The eight substantive cases in the book draw upon European and American thought and practice concerning 'the environment' in the centuries between 1500 and 1800. The collection 're-discovers' attributes about the idea of climate that had largely been erased by the hegemonic Earth sciences of the latter decades of the twentieth century but which are essential for understanding today's phenomenon of climate change. These include the observation that human knowledge about climate is more often 'bodily' and situated than it is detached and abstract; that both learned and popular discourses of climate and its changes co-exist in most cultures; that what is the 'problem' with a given climate is always contested; that human efforts to engineer the climate are usually exercises to control either communities or individuals; that climate is as much inside culture as culture is something which emerges from climate; and that the idea that *either* the environment 'rules' humans *or* humans 'rule' the environment is fallacious. In short, climate is bound up intimately with how people live and think and act. Climate is, and always has been, political.

We see this clearly with the Paris Agreement on Climate Change, signed in December 2015. This agreement represents the most inclusive and ambitious instrument of environmental governance, yet it was constructed and hailed at the time as a great triumph of international diplomacy. Its twenty-nine Articles were designed to "hold the increase in the global average temperature to well below 2°C above pre-industrial levels" and to promote further efforts to limit the increase to no more than 1.5°C. The agreement recognises the importance of all levels of government – national, regional, local – and also the roles of non-governmental political and social actors for securing this goal. The twenty-year negotiating process that led up to the Paris Agreement has normalised the idea that not only *can* global climate be governed, but it *must* be governed to avoid undesirable climatic outcomes.

But the emergence in recent decades of *global* climate as an object of governance redefines the nature and scope of the political. The Paris Agreement problematises climate in a very particular way: through global temperature. A rise of 2°C or more above pre-industrial temperature is deemed by the world's governments to be dangerous. But global temperature is not an object which is directly tractable through human actions. Governing global temperature therefore requires, at the least, governing the full range of human activities, technologies and institutions – and the imaginations which give rise to them – that emit greenhouse gases and other particulates into the atmosphere. This in turn requires virtually every human practice becoming subject, at least in principle, to the logic of global climate governance. Land, energy, mobility, diet, the academy, forests, procreation, design and, ultimately, all human behaviour become subject to the totalising idea of climate governance. Governing global climate becomes an exercise in governing global society.

The insights into the nature of climate and its politics that Miglietti and Morgan's book offers are increasingly recognisable today, as the hegemony of scientific knowledge about the environment has been questioned and as the meaning of climate change has been opened up to multiple and conflicting voices. Central to this necessary task has been the humanities, whose stock within the science of global environmental change has been rapidly rising. The value of the cases brought to our attention through historians in this book is that they show how thought and practice about climate have always been thus, at least within the North Atlantic societies and cultures of the early modern period. (There are other stories and histories yet to be told from outside this particular region, as the editors acknowledge.) Sometimes it is necessary to travel to a foreign country, where things are done differently, to see more clearly how things are done at home. By taking its readers to the marshes of seventeenth-century France, to the imperial Arctic projects of early-nineteenth-century Britain, to the wetlands of the newly independent east-coast states of north America and to the sixteenth-century Spanish colonies of South America, *Governing the Environment in the Early Modern World* helps us to recognise some of the less visible implications of today's environmental governance projects.

Mike Hulme

King's College London, 26 September 2016

Acknowledgements

This volume originates from a conference held on 16 May 2015 at the University of Warwick ('Ruling Climate: The Theory and Practice of Environmental Governmentality, 1500–1800'), organised by the volume's editors and sponsored by Warwick's Humanities Research Centre, the Centre for the Study of the Renaissance, the Research Student Skills Programme and the British Society for the History of Science. Most of the essays published here are substantially revised versions of papers presented on or prepared for that occasion. We would like to thank Sue Dibben and Jayne Brown for their invaluable support before, throughout and after the original conference; Anthony Carlson, Rebecca Earle, David A. Lines, Franz Mauelshagen and Anya Zilberstein for reading and commenting on a first draft of the introduction; and Ralph Bauer, David Beck, Andrea Frisch, James R. Fleming, John Gascoigne, Ariel Hessayon, Philippa Hellawell, Vladimir Janković, Piet van Cruyningen and Sam White for their insight and encouragement at every stage in the preparation of this book. Thanks are also due to Helen Bell, Rebecca Brennan, Sheri Sipka, Khanam Virjee and Kelly Watkins for their patient assistance throughout the editorial process.

Introduction

Ruling 'climates' in the early modern world

Sara Miglietti and John Morgan

How did people in the early modern world conceive of and attempt to leverage the relationship between themselves and their environments? The essays collected here shed light on this fundamental question through an interdisciplinary approach that takes into account both concrete environmental practices and the medical, philosophical, and political conceptions that sustained them. In so doing, this volume collectively pursues three main objectives: first, to bring together the all-too-often separate fields of intellectual history and environmental history in order to examine how early modern environmental governance functioned both in 'theory' and in 'practice'; second, to further our knowledge of early modern environmental practices and their social and political implications; and third, to demonstrate that early modern understandings of the environment were much more complex, contested, and ideologically invested than historians have hitherto given them credit for.

The relationship between environmental conditions and social and cultural formations in the early modern period has been the object of a number of recent studies. Historians have placed nature as a 'historical protagonist' at the centre of their narratives, demonstrating the relationships between a variety of phenomena from climatic change to global wind systems and the fates of states and empires.[1] In these studies, the environmental is analysed as an influence on the political. The essays in this volume similarly attend to the relationship between environmental conditions and political formations. Where they differ from this earlier work is that rather than analysing the historical link between environment and government, they examine how both single individuals and larger collectivities conceptualised this link and exploited it towards specific goals. In so doing, *Governing the Environment in the Early Modern World* seeks to reassess the relationship between environments and societies from the agents' own perspective, that is, by taking into account how early modern people themselves made sense of their relationship with their living milieus.[2]

Such an approach – pioneered in important works such as William Cronon's *Changes in the Land* (1983) and Richard Grove's *Green Imperialism* (1995) – is useful for various reasons, not least because it can help us avoid projecting our own modern categories onto the environmental conduct of past societies.[3] In this respect, the project of an environmental history written from the agents' perspective does present some challenges. First of all, the interventions into the physical world that we might term 'environmental' today were not necessarily considered to be such in the early modern period; it is not sufficient to consider forestry ordinances or river regulation as 'environmental' simply because they deal with aspects of what we consider 'the environment' today. Not all 'environmental regulation' was undertaken with specific environmental goals in mind. For example, in early modern Germany, village bylaws restricting access to various natural resources did not reflect an ecological or sustainable sensibility in their authors (the very terms 'ecology' and 'sustainability' being, of course, of a rather more recent coinage).[4] Rather, they were to guarantee harmony amongst neighbours co-habiting within fixed biological limits; their function was overridingly social rather than ecological in a modern sense.[5] Actions such as these are environmental only insofar as, to put it with one environmental historian, all our actions are environmental.[6] As historians, we might then study the environmental consequences of, say, state forestry policy, but we would be wrong to see such policies as being particularly preoccupied with an 'environment' that is rooted in modern conceptions of ecology.

It is then crucial to understand what, exactly, was the environment over which early modern societies sought to govern – if one existed at all. The essays presented here engage with this core question in various ways. What emerges overall is that among early modern Europeans there was no single notion of 'the environment' as we conceptualise it today. While the term itself made its first appearance in European vernacular languages as early as the thirteenth century,[7] it was long before it came to be used in the sense that we most frequently attach to it today – namely that of 'a set of interdependencies between Earth, animals, plants, climate and humans' that can be studied on either a local or a global scale. In Philemon Holland's 1603 translation of Plutarch's *Moralia*, the term is used alongside the English neologism 'circumplexion' (an encompassing of one thing by another) to translate the Greek *perieleusis* within a discussion of magnetic attraction that bears little connection to modern 'environments'.[8]

It is true that, at various times before the modern age, other terms were occasionally employed to refer to the ensemble of physical factors – primarily, though not exclusively, natural – that formed the context for all human activity on earth. In his *Tetrabiblos*, written in the second century AD, the Greco-Egyptian geographer and astrologer Claudius Ptolemy spoke of the climatic conditions of a given region in terms of *ta tou periechontos katastēmata* (literally, 'the conditions of what is all around').[9] But if Ptolemy's notion of *periechon* ('what is all around') may seem to come close to the modern meaning of 'environment', it did not quite cover the whole spectrum of physical factors that we now recognise as part of 'the environment'. Ptolemy was primarily interested in temperature and humidity, in keeping with Greek physics and meteorology, which were based on interactions between four fundamental elements or qualities: hot, cold, moistness, and dryness. He was also interested in astral influences, which he saw as the 'general causes' producing climatic diversity on earth as well as affecting human character and behaviour.[10] He was far less interested in examining the role that landscape features such as mountains or hills, soil quality, and the presence of running or standing waters played in shaping the climatic profile of a given place and its effects on local residents – although he did recognise at one point that 'even within the regions that in general are reckoned as hot, cold, or temperate, certain localities and countries have special peculiarities of excess or deficiency by reason of their situation, height, lowness, or adjacency'.[11]

Ptolemy's astro-meteorological ethnology, which featured a rough division of the inhabited world into four quarters dominated by different combinations of elemental qualities and astral influences (including those of the sun and the moon), would long remain an authoritative model for conceptualising environmental conditions and their effects on humans.[12] It was not, however, the only model available. In antiquity as in the early modern period, a range of different concepts were mobilised to describe the essential physical characteristics of a defined place, also with respect to their effects on local human populations. While a strand of early modern geographical thought looked back to classical theories of *klima* (literally 'slope' or 'inclination') to develop a tripartite view of the world as comprised of a hot zone, a cold zone, and a temperate zone in between,[13] other traditions similarly arising from classical antiquity worked in concert, and occasionally interfered, with this latitude-based outlook. For instance, the Hippocratic vocabulary of 'airs, waters, and places', which permeates early modern descriptions of urban and rural areas, draws attention to both meteorological factors (such as wind and rainfall) and specific landscape features (such as marshes, rivers, and mountains) as crucial components of what we would now call the 'climate' of a place and its overall ecological profile.[14]

Latitude, weather, air quality, and landscape features were thus seen as distinct and not always necessarily interrelated aspects of the physical milieus that human beings inhabited. Today we tend to bundle these concepts together and dub the result 'the environment' – a term that itself lacks a precise definition and is often imbued with shifting meanings depending on how and where it is used.[15] Early modern Europeans lacked such a unified concept of 'the environment'; their understandings of the physical world around them were irreducibly plural and varied greatly depending on the scale of the analysis and the explanatory model adopted.

While it is important to acknowledge such semantic gaps between modern and early modern notions of 'the environment', the term, we believe, can still be fruitfully applied to the study of early modern ideas and practices provided that we exercise some caution in utilising it. Throughout this collection, we propose to use the term 'environment' broadly to signify the spaces in which people lived and with which they engaged physically, intellectually, spiritually, and imaginatively. Following the insight of several recent commentators, we deem it especially important to understand the environment as the open-ended result of humanity's agency (be it creative or destructive, conscious or unconscious) rather than as a pre-existing given. Although we use the singular 'environment', it is important to recognise there is no single 'environment', instead there are many – always contextual and always historical. More particularly, we take 'the environment' to be the outcome of practices of 'environing' that produce environment*s* in the plural, that is, spaces 'where humans live and where humans have entered into a self-conscious relationship with their surroundings'.[16] Encompassing both the cultural process whereby humans perceive and conceptualise environments in their imagination and the physical process of 'place-making' whereby they concretely transform and reshape them, the concept of 'environing' allows us to think at once about the ways in which people have altered the world around them and about 'the symbolic transformations which configure "the environment" as a space for human action'.[17] There is an essential subjectivity in the shorthand 'the environment' that the chapters in this volume reveal in a variety of historical contexts.

The notion of environing is particularly useful when thinking about the relationship between environments and political interpretation and action. By attending to the ways in which people have historically produced their environments, we must retain a multiple focus. Environments should be understood in terms of their physical actuality, the intentions of their many creators, and the representations of those who encountered them. Environments are 'nature made social', sometimes even directed and shaped by understandings of the social and 'created as a solution to problems felt in societies'.[18] Because of this tight connection between environments and the societies that produce them, environmental discourse has often historically been bound up with the pursuit of particular social policies or political ideologies. The early modern period is no exception, and the way in which environmental theories have from time to time been bent to serve ideological and political agendas is at the centre of several contributions in this collection, including those of Dodsworth, Morera, and Zilberstein, which explore the issue over a broad chronological and geographical span. Other chapters in this volume address various aspects of early modern environings in relation to social and political practice, discussing attempts to control and manipulate the physical world, the motives behind these interventions, and how people interpreted the ways in which the physical might or might not manifest the political.

As products of a dynamic relationship between humans and nature, environments are to be understood as objects that are at once physical and cultural – at once given and constantly (re-)created. Environments, that is, are as much *inside* culture as *outside* of it, in that they constitute a precondition for all human experience as well as an ever-changing effect of it. In light of this fact, we believe that the joint study of environmental practices and ideas in context can provide a fresh approach to the environmental history of the early modern period.[19] The essays in this volume offer a few examples of how such an approach can be fruitfully applied to various geographical and historical contexts, including France, England, and the colonial Atlantic.

If environments can be interrogated as dynamic relationships between people and the world around them, the cultural constructions that sustain historical understandings of this relationship are then equally plausibly understood and interrogated as environmental. Whilst this volume deals with explicit attitudes and actions towards the environment, its aim is not to limit environmentally informed analyses of early modern politics to explicit statements of environmental ideas more generally. Other histories remain to be written that uncover the unspoken, assumed, and implicit environmental values on which aspects of early modern politics and statecraft rested.

Environmental influence, environmental governance, and climate change

Among the long-standing cultural constructions that early modern Europeans could draw upon to make sense of the relationship between themselves and their environments is a corpus of doctrines of ancient Greek origins, often known as 'climate theories' or 'theories of climatic influence'.[20] Though common in scholarly usage, such terms should be employed with a certain caution when referring to a period when modern scientific climatology did not yet exist, the causes of climatic diversity and heat distribution on earth were not yet clarified, and the word 'climate' itself still largely retained the technical cosmographic meaning of the Greek term *klima* from which it derived instead of being used in the modern meteorological sense.[21] 'Climate theory' is, in fact, a modern label, retrospectively imposed by scholars upon a diverse range of doctrines of environmental influence that were first advanced in classical Greece and then handed down to early modern Europe through a complex tradition.[22] Such doctrines ascribed a fundamental role to a number of environmental factors such as landscape, weather, and the stars in shaping the physical, mental, and moral constitution of human beings. From Hippocrates to Montesquieu, theorists working in this tradition construed such factors as major determinants of both individual and collective identity: it was primarily because of their different milieus – these theorists argued – that northern nations were rash, gluttonous, and skilled for all sorts of handicrafts, while people from hot places were prudent, vengeful, and adept in the speculative sciences.[23] Knowing the environmental components of 'national character' was thus of the utmost importance for understanding why certain peoples behaved in certain ways, how they should be governed (or how they could be conquered), and what was required to reform their character and conduct. As Jean Bodin argued in his 1576 *République*, effective statecraft implies 'accommodating' laws, policies, and political regimes to the environmental conditions of each particular country.[24]

As several chapters in this volume show, the link between environment and government was frequently somatic, that is, operated by (culturally charged) understandings of the body – both individual bodies and the collective body of the nation (the 'body politic').[25] If the environment was what people sought to influence or overcome, and people were what they sought influence over, then, as Rebecca Earle shows in her chapter, it was the body that they understood these influences to be working through. The significance of the relationship between the body, its environment, and order extended to the domestic realm. Sandra Cavallo and Tessa Storey have shown how ensuring 'good' airs in the domestic environment was a form of 'corporeal management' designed to promote good health.[26] Similar remarks have been made for urban spaces, increasingly subject in the early modern period to air and water pollution that contemporary residents often perceived as a danger for their physical and spiritual well-being.[27] In rural areas, the constitution of soils was linked to the character of local populations: soils were widely believed to be composed in the same way as human bodies, that is from combinations of four fundamental qualities, hot and dry or cold and moist, and to become 'out of heart' should these qualities become imbalanced.[28] Correspondingly, the drainage of the English fenland was claimed, by one anonymous poet, to have remarkable physical and moral effects on the humoural complexion of the fenlanders:

> When with the change of Elements, suddenly
> There shall a change of Men and Manners be;
> Hearts, thick and tough as Hydes, shall feel Remorse,
> And Souls of Sedge shall understand Discourse,
> New hands shall learn to Work, forget to Steal,
> New leggs shall go to Church, new knees shall kneel.[29]

In the sixteenth and seventeenth centuries, 'climate theory' preoccupied, among others, physicians, natural philosophers, geographers and cosmographers, religious missionaries, diplomats, and political theorists. From the latter half of the seventeenth century onwards, the wealth of knowledge collected through centuries of reflection on this matter was carefully vetted, classified, and digested in light of rising scientific epistemologies, ultimately coming to inform actual practices and policies on an unprecedented scale. In an age of growing concern of nation states with the strength and prosperity of their populations, 'climate theory' provided a long-standing, authoritative body of conceptual tools to think about national improvement, in both quantitative (e.g. demographic growth, increased productivity) and qualitative terms (e.g. physical health, military prowess, moral and civil conduct, intellectual skills).[30] It is therefore hardly surprising that European thinkers and policy makers would grow increasingly interested in exploring the dynamics of environmental influence and in developing strategies for coping with it. Diet, geographical mobility, and environmental engineering were among the most common 'ways of coping' with environmental influence theorised and practised throughout the early modern centuries, with environmental engineering rising to an ever-greater importance as we move through the period.[31] Several chapters in this volume reconstruct this history through case studies of early modern France, England, and the Americas from the early sixteenth to the early nineteenth centuries, thus shedding light on the considerable changes that occurred over this time span (see Earle, Spavin, Morera, Carlson). By the end of the period under consideration, we find traces of a remarkable paradigm shift in Jean-Baptiste Moheau's *Recherches et considérations sur la population de France* (Paris 1778), in which environmental influence is turned from a threatening, unmasterable force into a manageable tool of bio-political governance:

> It is up to the government to change the air temperature and to improve the climate; a direction given to stagnant water, forests planted or burnt down,

> mountains destroyed by time or by the continual cultivation of their surface, create a new soil and a new climate. [. . .] If the unknown principle that forms the character and the mind is the outcome of the climate, the regime, the customs, and the habit of certain actions, we can say that sovereigns [. . .] govern the physical and moral existence of their subjects. Perhaps one day we will be able to call on these means to give whatever hue we wish to morality and the national spirit.[32]

Moheau's words point to an important development in the historical understanding of climate. Classic theories of environmental influence had posited a link between national character and environmental conditions, naturally prompting the idea that by controlling not only space but place – namely, the physical landscape of a country and its peculiar set of environmental influences – one could also control those who inhabited it. The idea that humans could, to some extent, negotiate environmental influence by modifying their living milieus was an ancient one, which nourished a substantive literature on domestic economy, architecture, farming, and landscaping. The scale and significance of such transformations, however, was usually considered to be rather small. Very few believed that human agency could go so far as to permanently change the climate of a whole region. The situation changed dramatically in the early modern period, when the traditional view of climate as fixed and unalterable came to be replaced by one of climate as the shifting outcome of land-use patterns and other forms of human environmental agency. This conceptual revolution happened gradually and was not without complexity or resistance; yet although it would be mistaken to read too much continuity into the long genealogy of modern notions of anthropogenic climate change (studied by Zilberstein in this collection), it is indeed in this direction that early modern theories appear to have been moving.[33] Both environment and climate thus became increasingly politicised notions as they were placed at the centre of new power technologies that sought to govern people indirectly by manipulating their living milieus.[34] While the emergence of 'mesopolitics' – the art of governing (through) the milieu – is often traced back to the post-Enlightenment period,[35] the essays in this volume show that the origins of this new form of political rationality are actually to be sought in sixteenth- and seventeenth century theories of environmental influence and the governmental practices to which they gave rise.[36]

Drawing attention to the various ways in which 'climate theories' were mobilised in the early modern world in order to inspire, support, or justify such practices, the present collection sheds light on questions that have thus far remained in the margin of environmental historical studies: in a time in which the environment was thought to shape the physical, mental, and moral makeup of human beings, how was the relationship between environment and government conceptualised and exploited in specific historical and geographical contexts? How exactly was the environment thought to 'rule' over humans, and how did humans come to claim for themselves the power to 'rule' over the environment in turn? How and when did Europeans begin to conceptualise this power in 'mesopolitical' terms, as a means of governing humans *through* their climates and living milieus?

In order to answer these questions, the chapters in this volume bring together a diverse range of sources from landmark works of 'climate theory' (such as Jean Bodin's aforementioned *République*) to local administrative records, court proceedings, and radical political texts, all of which provide a stronger grasp of the various ways in which people understood and negotiated their relationship to their environments.[37] In using these documents to access a level of early modern environmental mentalities that tends to remain peripheral to intellectual histories of environmental thought, we follow Peter Coates's remark that understandings of the natural world can be engendered as much through labour and affective relationships in and with the environment as through the 'trickle down' of Big Ideas.[38] In an age in which agriculture and manual labour were central to everyday life, 'bodily knowledge' as much as scholarly knowledge played an important role in shaping the perception and understanding of one's environment.[39] Examining how 'learned' environmental discourse differed from but also interacted with 'popular' understandings of the relationship between people and their environments is crucial for a fuller appreciation of early modern environmental ideas and their practical applications in various specific contexts.

This approach also sheds light on the different pace at which environmental ideas and environmental practices evolved, with practice sometimes leading developments in theory and sometimes following in their wake. The early modern period was a time of rapid intellectual change on many fronts, with critical breakthroughs in disciplines such as medicine, astronomy, and 'natural philosophy', as the natural sciences were then called.[40] However, the pace of these transformations did not always coincide exactly with the pace at which their effects began to be felt more broadly in the cultural and practical life of European societies. As the long histories of 'defeated' scientific paradigms such as astrology and humoural medicine (on which see Earle and Carlson in this volume) demonstrate, it often takes time for novel ideas to gain wide acceptance, dispel old paradigms, and filter down into actual practice.[41] Quite frequently, intellectual change comes not in the form of conceptual revolutions but in that of traditional theories that are adapted and transformed as they are used (and misused, and abused) in real-life contexts. 'Learned' and 'popular' environmental discourses should therefore be studied together if we intend to clarify not only the ways in which environmental theory and environmental practice interacted in the early modern period but also how environmental theories themselves evolved.

The transition from theory to actual policy often involved the identification of certain aspects of particular environments as problematic. The increased identification of environmental 'problems' is indicative of the new governmental approach to the environment in this period: in nature there are no 'problems'; they exist only in environments, which themselves have been environed – that is, physically and culturally inhabited – by man.[42] Several essays in this volume show how aspects of these environments were 'rendered thinkable in such a way as to be practicable or operable' by early modern political actors.[43] From the sixteenth century onwards, the early modern period witnessed an unprecedented degree of intervention into and engineering of the physical world for economic and political gain. Dramatic landscape transformation took place across much of Europe and the colonised world throughout this period, as new rivers were cut, swamps drained, land reclaimed, forests cleared, land irrigated, and commons enclosed. Some of these transformations have been studied from social, cultural, and legal perspectives, but their scientific and intellectual foundations still warrant further examination. For instance, the process of drainage and enclosure in England has generated a significant amount of scholarship that has focussed on the economic rights enjoyed and then lost by commoners and the concomitant political and cultural foment that accompanied this transformation.[44] The most innovative studies have considered the eco-legal aspects of enclosure and have characterised the double-quickset black and white thorn hedgerows used by enclosers as a kind of 'organic barbed wire'.[45] Early modern landscapes in general were inscribed with new legal topographies that reshaped how the environment was experienced. The essays in this collection shed light on the cultural and scientific premises of these practical transformations as well as on the repercussions that the latter had on further scientific change.

Deliberate, large-scale landscape change was nothing new; many of these trends had historical precedent in both the immediate and distant past. The medieval centuries witnessed significant periods of deliberate landscape change, some of which was aimed at altering local environmental conditions. Religious houses had intervened in their environments for spiritual and territorial reasons for centuries. Attracted by the positive religious virtue attached to remoteness, early Christians sought out woods and wetlands to found their religious houses. The abbeys of Crowland and Glastonbury were founded after the drainage of large expanses of marshland undertaken as a religious exercise.[46] The inhabitants of Languedoc-Roussillon drained and subsequently irrigated wetlands in Montady to produce a landscape suitable for supporting arable farming in the thirteenth century.[47] Urban communities throughout England had proved themselves watchful managers of water, waste, and smoke since the thirteenth century and continued these regulatory endeavours into the early modern period.[48]

Despite the scale of medieval landscape management and manipulation, that which was undertaken in the early modern period differed in both significance and degree. While religious motives remained an important source of legitimisation for landscape alteration schemes,[49] the early modern period also witnessed the emergence of new forms of environmental governance for politico-economical purposes. In England and France, the subject of Raphaël Morera's chapter, state-sanctioned drainage projects were undertaken throughout the seventeenth century in an attempt to improve the economic viability, climate, and governance of wetland areas. European monarchs strengthened the organisation of water management in the sixteenth century, empowering landowners to embank and defend flood-liable land through a series of legal reforms. Responding to increased population density and the more frequent incidence of disastrous fires, urban governments introduced fire regulations and building reforms in an attempt to manage one of the most profound threats to life in early modern towns.[50] Forestry became a central concern of European states, as they sought to protect a central natural resource of the 'wooden age': as the early-seventeenth-century English commentator Arthur Standish surmised, 'no wood, no kingdome'.[51] On a smaller scale, farmers began to apply new intensive agricultural techniques that required more precise and frequent manipulation of their natural resources. Water meadows, field-scale drainage, and the embankment of intertidal land from the sea became increasingly common from the early seventeenth century. Early modern governments and landowners became increasingly interested in the function and form of a variety of different ecologies and landscape types and saw altering them as a viable way of influencing the populations over which they governed.

These new abilities and desires to control the physical world for gain were summed up in microcosm in the garden. Gardens based on geometric plans and containing species collected from across the world appeared across early modern Europe, including in Padua, Leiden, Oxford, and Paris. These gardens, according to Carolyn Merchant, 'symbolized both an improvement of nature through labour and an improvement of the human condition'.[52] Cressy Dymock proposed a large-scale vision of the farm-as-garden in 1653, designing the layout of the 'little world' of the farm so as to keep people, plants, animals, and water productive and under control.[53] With these changes, Dymock argued, England would no longer need to rely on other nations and stood to gain politically and economically from the consolidation and schematisation of its farmland.[54] Contributions to this volume analyse large-scale environmental projects such as these in England, North America, South America, and France.

Frequently, practical and economic literature formed the most explicit link between environmental theory and political practice. The ideas of the natural philosophers and their predecessors are found in more concise, practical forms in writers with more immediately instrumental practical and political aims. In England, a vast literature of 'improvement' was generated by men who sought to derive specific material and political benefits through a variety of projects.[55] In Germany, *Hausväterliteratur* considered similar methods by which estates might be more profitably disposed of through the patriarchal rule of the household, new stewardship practices, and the readoption of classical agricultural methods from writers such as Columella and Cato.[56] Proponents of these new agricultural techniques wrote about them in moralistic tones that suggested and sometimes directly claimed that the character of peoples and nations would be improved through these works. In his *English Improver* (1649), Walter Blith urged Parliament to encourage improvement, as 'Mens spirits will be raised to such Experimenting of the Principles of Ingenuity, as that we may see this Kingdome soone raised to her utmost fruitfulnesse and greatest glory.'[57] Blith's optimism was reflected in enabling legislation passed by governments seeking the advantages of improvement. In England, drainage acts in 1600 and 1649 referred to the 'great and inestimable benefite' arising to 'Bodies Politique [and] Corporate' from draining 'wastes' and common marshes and later to the 'great advantage and strengthening of the Nation' that would come through drainage.[58] Such improvements were channelled through the body. In 1700, Timothy Nourse, a particularly enthusiastic improver, claimed that labouring on improvement works 'quickens Appetite, and contributes to Health and Strength of Body'.[59]

Such projects were not restricted to the so-called Old World. As this collection shows, discussions spread across Europe and the wider world as to how to improve land and climate for human benefit. Encounters with entirely new contexts in colonial settings and changing local environments in Europe led to the reconsideration of long-held beliefs about the role of climate in upholding the social order, driving economies, and affecting public health. Everywhere European colonists went they observed and directed changes in local environments. In the New World, colonists observed the rapid environmental changes that came with the extractive practices of colonial enterprises. Spanish colonists remarked on the changes they saw occurring in the new lands they encountered: in 1579 Diego de Esquivel noted that, since the arrival of the Spanish, the people of Chinantla had declined in numbers, reducing the ground under tillage and restricting their ability to clear jungle, leading to an increase in swamps, humidity, and ultimately ill health.[60] Colonial powers identified and attempted to solve environmental 'problems' they encountered in new lands: the Spanish undertook an enormous drainage enterprise to stop what they saw as problem flooding in colonial Mexico City, and in the Transkei region of what is now modern South Africa, changes Dutch settlers made to agricultural practices turned fire from a nutrient-cycling resource into a threat that required containing.[61] The problems colonial powers encountered were frequently solved with colonial knowledge and local expertise; in Mexico City the Spanish utilised indigenous hydraulic technology, and in Ottoman Egypt, Ottoman governors relied heavily on local expertise to manage irrigation.[62]

Whilst sixteenth- and seventeenth-century colonial domination was characterised by extractive environmental activities, colonists shifted the focus of their control to a more reflexive style of environmental management in later centuries. During the eighteenth century, island colonies such as St Helena and Mauritius were subject to attempts to control the environment in response to the degradation colonial powers had inflicted throughout the later seventeenth century. On Mauritius, for example, French colonial governors sought to mitigate and reverse the impacts of the colonial slave-plantation economy through the imposition of conservationist measures aimed at regulating the cutting of wood and the flow of canals.[63] These few examples and the papers on travel and colonialism, imperial science, and North American drainage in this volume (by Earle, Zilberstein, and Carlson, respectively) indicate that any comprehensive study of environmental governance across the early modern period requires a transnational perspective to fully understand how political engagements of the environment developed in the long term, under the influence of new ideas and new experiences.

Overview of the chapters

The chapters in this volume go beyond Europe, but they do not go beyond Europeans. Nor does *Governing the Environment* cover Africa, Asia, Eastern Europe, the Middle East, or Oceania. The omission of these areas is not due to their irrelevance. Important environmental histories of governance have already been written about some of these regions: Richard Grove's seminal *Green Imperialism* brought the activities of the English and Dutch East India Companies in Asia, the Caribbean, and the Pacific and Indian Oceans to the centre of historical narratives about environment and government in the extra-European world.[64] Alan Mikhail and Sam White have set the agenda for Ottoman environmental history with two important books on water and state power and climate and rebellion in the early modern Ottoman empire.[65] Wenkai He has shown how the Qing state expanded its reach into local society by utilising a discourse of 'public goods' to intervene in non-state hydraulic projects.[66] The chapters assembled here proceed in a rough chronological order to address the interventions of the English, French, Spanish, Portuguese, and North Americans, both at home and abroad.

Rebecca Earle's chapter, on climate, travel, and colonialism in the early modern world, ties together several key themes running throughout the volume. Looking primarily at Spanish American colonists, Earle asks, how did colonists understand and attempt to mitigate the effects of living in foreign and often unfamiliar environments? Earle finds that Spanish colonists interpreted the novelty of New World environments through classical humoural theory. Early modern travel writers warned potential voyagers of the dangers that new environments posed to the humoural balance of their bodies. As Earle's chapter shows, the body, as the medium through which environments worked on people, was a central concern in early modern environmental thought. Unfamiliar environments could have serious and debilitating impacts on both physical health and emotional state through the body's ill suitedness to local climates, soils, and foods. Earle shows how temperament was affected by environments and, ultimately, how the fate of the Spanish colonial endeavour rested in some way on large-scale attempts to cope with new environments. More often than not, this coping involved significant amounts of environmental change. Colonists imported basic old-world commodities like red wine, olive oil, and wheat in order to help their bodies cope with the upheaval. But they also began to make changes to the landscapes they inhabited, growing wheat, olives, and grapes. Thus quotidian aspects of European life became building blocks for Spanish life in the Americas. American environments needed to be able to reproduce European agriculture and European commodities in order to sustain a successful colony.

Richard Spavin's chapter addresses this relationship between governing regimes and their environments through a re-reading of Jean Bodin's *Six Books of the Commonwealth*. In spite of Lucien Febvre's protestations in his *Geographical Introduction to History*, Bodin has been characterised as one of the clearest exponents of a crude kind of environmental determinism.[67] He has been enlisted, along with Montesquieu, as one of two prescriptive, deterministic, and classically influenced bookends for early modern environmental thought.[68] In Bodin, Spavin identifies the concept of 'ana*chorism*' – of deliberate geographical inaccuracy – to reinterpret this legacy. Reading Bodin within the context of the religious and political turmoils of sixteenth-century France, Spavin excavates several formulations of 'climate theory' from across Bodin's writings. By analysing the inconsistencies between Bodin's various formulations of the relationship between climate, geography, and human society, Spavin questions Bodin's legacy as an environmental determinist and even as a climate theorist. Instead, Spavin argues, Bodin ought to be seen as creatively engaging a corpus of classical ideas at specific moments across the sixteenth century in constant intertextual dialogue with his own work and that of other writers. These intellectual and literary contexts enable us to see Bodin's use of 'climate theory' as less an endorsement of classical ideas and more a comment on contemporary France. By replacing Aristotelian 'hellenocentrism' – the belief that Greece had the optimum climate for an advanced civilisation – with his own 'gallocentrism', Bodin could explore the 'geographical errors' of French political and religious turmoil.

Raphaël Morera examines how Bodin's writings related to practical political projects in the seventeenth century. In his chapter on marshes as 'microclimates', Morera traces how humoural and climatic theories found in Bodin and other Renaissance authors were deployed in a number of drainage schemes on French marshlands. Morera shows how the French monarchy became increasingly interested in environmental regulation during the early modern period. Edicts and new legislation enabled the better administration and closer oversight of rivers, marshes, and forests. Marshes and wetlands in particular were seen as underexploited regions. Their supposedly bad airs and unenclosable, uncultivable landscapes were portrayed as wastelands and the people that inhabited them as obstinately backward, economically unproductive, and politically ungovernable. As Morera shows, the monarchical impetus for drainage was then both economic and political: to drain was to gain productive land from unproductive waste, 'improve' the resident population, and extend the reach of government. Morera shows how local environmental conditions were dramatically changed at the behest of an expanding state and in response to classically inflected 'climate theories'. In seventeenth-century France, marshes were symbolically significant – uncultivated, unimproved land was a sign of a lack of civilisation and political order. Drainage, in the monarchical ideology described by Morera, served to integrate marginal areas within the broader polity by erasing climatic aberrations and reconciling them with the 'true' French environment.

In her paper on north-east England's River Tyne, Leona Skelton also analyses the ideas and goals that lay beneath practical environmental management. Bringing new sources and new voices into the conversation about early modern environments, Skelton demonstrates how records from the archives of social history can be applied to further our understanding of early modern environmental management. Skelton uses the minutes of the Tyne River Court – which was appointed to regulate the Tyne estuary in 1613 by the river's official conservators, Newcastle Corporation – to reveal a long history of conservation and management of the River Tyne. Records of local administration remain under-exploited by early modern environmental historians. Medievalists have made good use of court records to reveal environmental histories of disease and urban farming and in climatic reconstruction.[69] Yet most early modern historians have viewed courts and administrative bodies as social institutions pursuing social ends.[70] As Skelton demonstrates, they contain significant evidence of environmental attitudes, policies, and objectives. Looking at the regulations enforced on the Tyne between the mid-sixteenth and the nineteenth centuries, Skelton carefully reconstructs the changing ways in which administrators in Newcastle viewed and interacted with their naturally powerful river, on which they relied so heavily for their income. Skelton shows that through frequent, regular, and practical engagements with the Tyne, its conservators developed complex and empirically grounded methods for controlling the impacts of external environmental phenomena and thus protecting the river, the economic heart of a city and a region.

In like manner to Skelton, Ashley Dodsworth takes up sources well used by social and political historians – the writings of seventeenth-century English radical Gerrard Winstanley – and asks how we might understand them as environmental in her chapter on the importance of land and labour to Winstanley's politics. As Donald Worster has shown, environmental history can be 'done' in any number of places, from the field to the factory, it just takes critical effort from us as historians to recognise the environmental.[71] Likewise, we can excavate the environmental from the sources we are most comfortable with as early modernists. In her contribution, Dodsworth provides an environmental re-reading of Winstanley's political writings. Focussing on the importance Winstanley attached to labour, Dodsworth shows that ownership of, access to, and cultivation of the land was integral to Winstanley's radical vision for England in the 1640s and 1650s. For Dodsworth, the key to Winstanley's thought is his understanding of governance as an embodied phenomenon: that it is acted and experienced through affective relationships with the physical environment. Labouring on land that was held privately, in an environment that was 'locked up', contributed to the subjugation of the English peasantry. For Winstanley, liberation would be achieved through a radical opening of natural resources, in which labour on free land was the only way to ensure a free and just society. By paying attention to the importance of labour and the transformation of 'waste' land, Dodsworth reveals the environmental – but by no means 'green' – aspects of Winstanley's thought.

William Cavert addresses another aspect of the relationship between environmental and political order in the early modern period. Analysing a number of responses to harsh winters and extreme cold in early modern England, Cavert reassesses the long-held historiographical assumption that in the 'enchanted', Divinely malleable, and providential world of seventeenth-century England, changes in the weather were understood as judgements on the rulers of the day. Focussing on responses to the unusually cold winters of 1607/8 and 1684, Cavert's chapter nuances recent discussions of weather events as Divine portents and judgements on contemporary events by revealing a diversity of responses to harsh winters. Cavert shows that instead of inducing crises, freezing conditions provided opportunities for sociability, mirth and play at ice fairs, observation and discussion among newsletter correspondents (little of which was political), and moments of state formation rather than disintegration through opportunities for charity and relief provision. This history of politically uncontentious hard winters is important. In an era in which environmental phenomena were invested with so many political meanings and interpretations, Cavert shows we cannot take such responses for granted. It is testament to the diversity of ideas about the role of the environment in social and political life that at moments in which we might expect the environment to play a starring role in explanations of politics, it is conspicuously absent. To invoke the relationship between environment and society was to make claims about the world that were frequently political in nature. The lack of politics attached to events we might see as prime for political interpretation de-couples early modern environmental thought from early modern political thought. Cavert's analysis helps us see that environmental influence was not taken as a given; those who invoked it were not necessarily parroting long-established commonplaces, they were doing intellectual work.

Anya Zilberstein looks at another instance of political readings of environmental phenomena in her chapter on the transmission of knowledge about glacial retreat in the scientific networks of Sir Joseph Banks (1743–1820). Banks, the most influential scientific patron in the English-speaking world, directed much of the scientific investments made across Britain's empire. From his position as head of the Royal Society, in the heart of the metropole, Banks corresponded with a wealth of scientific observers, offering description and analysis of phenomena from across the world. In his various positions at the heart of the British state – the Colonial and Home Offices, the Admiralty, and the Board of Agriculture – Banks retained quite different interests. Zilberstein demonstrates how Banks's multiple public roles coalesced in his sponsorship of expeditions to the Arctic in search of a northwest passage. Responding to news about climatic change and a potentially thawing north-western shipping route, Banks jumped on unreliable and unverified scientific observations to justify a project that would be of great importance to the empire were a passage to be found. Zilberstein shows us not only an early example of the politicisation of climate science but a clear example of the political nature of practices of environing. Banks's enthusiasm for the discovery of a northwest passage shows us the extent to which the climate had become an object of both scientific analysis and state interests by the early nineteenth century. Drawing on a long history of speculation about the nature of climate and driven by social, economic, and political rationale, Banks was prepared to imagine and invest in an Arctic that piqued both his scientific and political interests.

Anthony E. Carlson's chapter, along with Earle's and Dodsworth's, highlights the importance of the body in understandings of environmental influence and attempts to manipulate it for gain. Carlson demonstrates the continued importance of climatic, miasmic, and humoural theory in the United States in the later eighteenth and nineteenth centuries, showing how the spread of yellow fever in the 1790s brought these much older, European ideas to the fore in medical and agricultural discussions of public health and federal as well as state land-use policy. Discoveries in medicine, meteorology, and pneumatic chemistry pointed to water – particularly standing and stagnant water – as the source of miasmas, the cause of yellow fever, and thus the determinant factor in the creation of unhealthy climates. Tracing the influential discourses of humouralism and miasma through European writings on agricultural science, through local American agricultural magazines, and eventually into the official record of the US Congress, Carlson shows that Americans continued to grapple with, succumb to, and attempt to influence their climates and environments well into the nineteenth century. They did this through a series of landscape interventions by public drainage authorities, displaying a strong faith in public institutions' capability to intervene in local environments on public health grounds. For late-eighteenth and nineteenth-century Americans, disease linked people, environmental conditions, and place. Carlson's chapter demonstrates the enduring importance of notions of environmental influence and refutes the idea that, by the early nineteenth century, Americans had dismissed the influence of their surroundings and mentally conquered their climate.

With Carlson's essay we reach the threshold of the industrial age, a period in which the multiple interconnections of science, environmental governance, political ideology, and social engineering examined in this volume did not disappear but rather were strengthened and refined.[72] It remains to be seen to what extent these later developments stand in continuity with the earlier phenomena studied in this volume or whether changes in medical and scientific thought (including the decisive shift from a humoural and flexible to a biological and fixed view of the human body),[73] technological advances, and social and political transformations in the 'age of the masses' brought about significant changes in the theory and practice of 'mesopolitics'. But this is food for further thought. Our main ambition here has been to identify and test new avenues for research that may fruitfully be pursued by early modern scholars as well as by scholars in other periods.

Collectively, this collection raises more questions than we may have been able to answer; occasional dissonances may be spotted, arising from the different contexts examined, sources used, and methodologies employed; but all of this is testament, we believe, to the diversity of approaches that could and should go on to contribute to a long-term history of environmental theories and practices. As the following chapters show, the early modern period is an especially important, though lesser-known, chapter in the genealogy of Western debates over climate, climate change, and environmental governmentality. Strategically nested between the Renaissance and its recovery of classical knowledge on the one hand and the industrial age with its fast-paced social, political, and environmental transformations on the other, the early modern period occupies a central place in this centuries-long history and one that, as we hope to show, deserves further attention for its creative blending of old and new modes of understanding and engaging with climate and environmental problems. Our hope is that the present collection will pave the way to further interdisciplinary study of the long and complex history of environmental governance and of its political uses – a topic that we can only expect to become ever more central to the concerns of our own age.

Notes

1 Bruce M. S. Campbell, 'Nature as Historical Protagonist: Environment and Society in Pre-Industrial England', *The Economic History Review* 63, no. 2 (2010), 281–314; Greg Bankoff, 'Winds of Colonisation: The Meteorological Contours of Spain's Imperium in the Pacific 1521–1898', *Environment and History* 12, no. 1 (2006), 65–88; Geoffrey Parker, *Global Crisis: War, Climate Change and Catastrophe in the Seventeenth Century* (New Haven, CT: Yale University Press, 2013); Sam White, *The Climate of Rebellion in the Early Modern Ottoman Empire* (Cambridge: Cambridge University Press, 2013).

2 For the importance of an environmental history oriented towards the retrieval of the historical agents' meanings and intentions, see Christian Pfister, 'Weather, Climate, and the Environment', in *The Oxford Handbook of Early Modern European History, 1350–1750, Vol. 1: Peoples and Place*, ed. Hamish Scott (Oxford: Oxford University Press, 2015), 70. A similar approach has been attempted for the Enlightenment and post-Enlightenment periods by, for instance, Jan Golinski, *British Weather and the Climate of Enlightenment* (Chicago and London: University of Chicago Press, 2007) and *Klima*, ed. James R. Fleming and Vladimir Janković, special issue of *Osiris* 26, no. 1 (2011), 1–270.

3 William Cronon, *Changes in the Land: Indians, Colonists, and the Ecology of New England* (New York: Hill & Wang, 1983); Richard Grove, *Green Imperialism: Colonial Expansion, Tropical Island Edens and the Origins of Environmentalism, 1600–1860* (Cambridge: Cambridge University Press, 2003).

4 For a genealogy of the notion of 'sustainability', see Paul Warde, 'The Invention of Sustainability', *Modern Intellectual History* 8, no. 1 (2011): 153–170. Although the word 'ecology' is not attested until 1866, when it was first introduced by the German biologist Ernst Haeckel (1834–1919), Donald Worster has argued that ecological ideas *avant la lettre* can already be located in the eighteenth century (*Nature's Economy: A History of Ecological Ideas* [Cambridge: Cambridge University Press, 1994]).

5 Paul Warde, 'Imposition, Emulation and Adaptation: Regulatory Regimes in the Commons of Early Modern Germany', *Environment and History* 19, no. 3 (2013), 313–337.

6 Paul Sabin, 'Rooting around in Search of Causality', *Environmental History* 10, no. 1 (2005): 83.

7 Gary Wickham and Jo-Ann Goodie, *Legal and Political Challenges of Governing the Environment and Climate Change: Ruling Nature* (London and New York: Routledge, 2013), 42–43.

8 OED, 'Environment', n. 1. http://www.oed.com/view/Entry/63089. Last accessed 9 June 2016.

9 Claudius Ptolemy, *Tetrabiblos*, 2.2.58, ed. F. E. Robbins (Cambridge, MA: Harvard University Press, 1940), 127.

10 Ptolemy, *Tetrabiblos*, 2.1, 117. On ancient and early modern meteorology, see Craig Martin, *Renaissance Meteorology: From Pomponazzi to Descartes* (Baltimore, MD: Johns Hopkins University Press, 2011). For an ancient application of elemental meteorology to medical thought, see the collection of pseudo-Aristotelian *Problems* in *The Complete Works of Aristotle: The Revised Oxford Translation*, ed. Jonathan Barnes, 2 vols. (Princeton, NJ: Princeton University Press, 1984), 14.7.41–42, ii, 1414.

11 Ptolemy, *Tetrabiblos*, 2.2, 127.

12 Marian J. Tooley, 'Bodin and the Mediaeval Theory of Climate', *Speculum* 28, no. 1 (1953), 64–83.

13 For ancient theories of *klima*, see Alexander Altmann,'Judah Halevi's Theory of Climates', *Aleph* 5 (2005), 215–246. For their medieval and early modern reception, see N. Wey Gómez, *The Tropics of Empire: Why Columbus Sailed South to the Indies* (Cambridge, MA: The MIT Press, 2008).

14 Clarence Glacken, *Traces on the Rhodian Shore: Nature and Culture in Western Thought from Ancient Times to the End of the Eighteenth Century* (Berkeley: University of California Press, 1967), 80–115; Andrew Wear, 'Place, Health, and Disease: The Airs, Waters, Places Tradition in Early Modern England and North America', *Journal of Medieval and Early Modern Studies* 38, no. 3 (2008), 443–465. See below, section "Environmental influence, environmental governance, and climate change" in this Introduction.

15 Philip W. Sutton, *The Environment: A Sociological Introduction* (Cambridge: Polity Press, 2007), 1–4.

16 Sverker Sörlin and Paul Warde, 'Making the Environment Historical: An Introduction', in *Nature's End: History and the Environment*, ed. Sverker Sörlin and Paul Warde (Basingstoke: Palgrave Macmillan, 2009), 2–3; Paul Warde, 'The Environmental History of Pre-Industrial Agriculture in Europe', in *Nature's End*, ed. Sverker Sörlin and Paul Warde (Basingstoke: Palgrave Macmillan, 2009), 75; Sverker Sörlin, 'Reconfiguring Environmental Expertise', *Environmental Science & Policy* 28 (2013): 16; Dolly Jørgensen and Sverker Sörlin, 'Making the Action Visible: Making Environments in Northern Landscapes', in *Northscapes: History, Technology, and the Making of Northern Environments*, ed. Dolly Jørgensen and Sverker Sörlin (Vancouver: UBC Press, 2013), 5.

17 Hannes Bergthaller, Rob Emmett, Adeline Johns-Putra, Agnes Kneitz, Susanna Lidström, Shane McCorristine, Isabel Pérez Ramos, Dana Phillips, Kate Rigby and Libby Robin, 'Mapping Common Ground: Ecocriticism, Environmental History, and the Environmental Humanities', *Environmental Humanities* 5 (2014): 267. On 'place-making', see John Dixon Hunt, *Greater Perfections: The Practice of Garden Theory* (London: Thames and Hudson, 2000).

18 Sörlin, 'Reconfiguring Environmental Expertise', 16.

19 Vladimir Janković, *Confronting the Climate: British Airs and the Making of Environmental Medicine* (New York: Palgrave Macmillan, 2010), 4; Mary Floyd-Wilson, *English Ethnicity and Race in Early Modern Drama* (Cambridge: Cambridge University Press, 2003), 29.

20 Frank Lestringant, 'Europe et théorie des climats dans la seconde moitié du XVIe siècle', in *La conscience européenne au XVe et au XVIe siècle. Actes du colloque international organisé à l'Ecole Normale Supérieure des Jeunes Filles (30 sept.–3 oct. 1980)* (Paris: Ecole Normale Supérieure des Jeunes Filles, 1986), 206–226; Mario Pinna, 'Un Aperçu historique de "la théorie des climats"', *Annales de géographie* 547 (1989): 322–325; Reimar Müller, 'Montesquieu über Umwelt und Gesellschaft – die Klimatheorie und ihre Folgen', *Sitzungsberichte der Leibniz-Sozietät* 80 (2005): 19–32.

21 For a reappraisal of the conceptual history of climate, see Franz Mauelshagen, 'Ein neues Klima im 18. Jahrhundert', *Zeitschrift für Kulturwissenschaften* 1 (2016): 39–57.

22 See Glacken, *Traces on the Rhodian Shore*; Lucien Boia, *The Weather in the Imagination*, trans. Roger Leverdier (London: Reaktion Books, 2005); Sara Miglietti, 'Mastering the Climate: Theories of Environmental Influence in the Long Seventeenth Century' (PhD dissertation, University of Warwick, Coventry, 2016).

23 Montesquieu may at once be seen as the apogee of an older tradition and as the initiator of a new one. On his transitional role, see Mauelshagen, 'Ein neues Klima im 18. Jahrhundert'. On Montesquieu as a 'climate theorist', see, in this volume, Morera, 'Marshes as Microclimates' and Spavin, 'Jean Bodin and the Idea of Anachorism'.

24 Jean Bodin, *Les Six livres de la République* (Paris: Jacques Du Puys, 1576), 516 (5.1: 'Du reiglement qu'il faut tenir pour accommoder la forme de République à la diversité des hommes').

25 For this link between individual bodies and the body politic, and how it operates in early modern 'biopolitics', see David Glimp, *Increase and Multiply: Governing Cultural Reproduction in Early Modern England* (Minneapolis: University of Minnesota Press, 2003).

26 Sandra Cavallo and Tessa Storey, *Healthy Living in Late Renaissance Italy* (Oxford: Oxford University Press, 2013), 8; Sandra Cavallo, 'Health, Air and Material Culture in the Early Modern Italian Domestic Environment', *Social History of Medicine* 29, no. 4 (2016), 695–716.

27 Mark Jenner, 'The Politics of London Air: John Evelyn's *Fumifugium* and the Restoration', *Historical Journal* 38, no. 3 (1995), 535–551; *Curare la città: sanità e igiene a Firenze, Roma, Parigi, Londra, Barcellona*, ed. Renato Sansa, special issue of *Storia Urbana* 112 (2006); Ken Hiltner, *What Else Is Pastoral? Renaissance Literature and the Environment* (Ithaca and London: Cornell University Press, 2011), Chapter 5; William M. Cavert, *The Smoke of London: Energy and Environment in the Early Modern City* (Cambridge: Cambridge University Press, 2016).

28 Verena Winiwarter, 'The Art of Making the Earth Fruitful: Medieval and Early Modern Improvements in Soil Fertility', in *Ecologies and Economies in Medieval and Early Modern Europe*, ed. Scott G. Bruce (Leiden: Brill, 2010), 99; Warde, 'The Invention of Sustainability', 156.

29 Jonas Moore, *The History or Narrative of the Great Level of the Fenns, Called Bedford Level with a Large Map of the Said Level, as Drained, Surveyed, & Described by Sir Jonas Moore Knight, His Late Majesties Surveyor-General of His Ordnance* (London: Moses Pitt, 1685), 75–76. On the dispute surrounding these drainage schemes and the competing ideological arguments mobilised therein, see Hiltner, *What Else Is Pastoral?*, Chapter 6.

30 See Glacken, *Traces on the Rhodian Shore* (especially Chapters 1.2, 2.2, 3.2); Lestringant, 'Europe et théorie des climats dans la seconde moitié du XVIe siècle', 218–221; Floyd-Wilson, *English Ethnicity*; Glimp, *Increase and Multiply*; Boia, *The Weather in the Imagination*, 23.

31 See Miglietti, 'Mastering the Climate'.

32 'Ces moyens de diriger et de changer le cours de la population sont dans la main du gouvernement; sa puissance est plus étendue encore; souvent il dépend de lui de changer la température de l'air et d'améliorer le climat. Un cours donné aux eaux croupissantes, des forêts plantées ou brulées, des montagnes détruites par le temps ou par la culture continuelle de leur superficie, forment un sol et un climat nouveau. Si du climat, du régime, des usages, de l'habitude de certaines actions, il résulte le principe inconnu qui forme les caractères et les esprits, on peut dire que les Souverains [. . .] régissent l'existence physique et morale de leurs sujets. Peut-être un jour pourra-t-on tirer parti de ces moyens pour donner aux moeurs et à l'esprit de la nation une nuance à volonté' (Jean-Baptiste Moheau, *Recherches et considérations sur la population de France* [Paris: Moutard, 1778], 291–292; quoted and translated in Michel Foucault, *Security, Territory, Population: Lectures at the Collège de France 1977–1978*, trans. Graham Burchell (Basingstoke: Palgrave Macmillan, 2007), 22–23. For a discussion of this text, see Miglietti, 'Mastering the Climate', 278–281, 288–289.

33 James Fleming, *Historical Perspectives on Climate Change* (Oxford and New York: Oxford University Press, 1998).

34 Michel Foucault, *Naissance de la biopolitique: Cours au Collège de France (1978–1979)* (Paris: Gallimard, 2004). For a useful introduction to the notion of governmentality, see Margo Huxley, 'Geographies of Governmentality', in *Space, Knowledge, Power: Foucault and Geography*, ed. J. Crampton and S. Elden (London: Ashgate, 2007), 185–204; Peter Miller and Nikolas Rose, *Governing the Present: Administering Economic, Social and Personal Life* (London: Polity Press, 2008).

35 On mesopolitics ('mésopolitique') see Ferhat Taylan, 'La Rationalité mésologique. Connaissance et gouvernement des milieux de vie (1750–1900)' (PhD Thesis, Université de Bordeaux III, Bordeaux, 2014).

36 According to Mark Harrison, *Climates and Constitutions: Health, Race, Environment and British Imperialism in India, 1600–1850* (Oxford: Oxford University Press, 1999) and Jean-Baptiste Fressoz, *L'Apocalypse joyeuse. Une histoire du risque technologique* (Paris: Seuil, 2012), among others, it was only after 1800 that attitudes towards environmental influence shifted from being predominantly adaptive or reactive towards being increasingly

proactive, as the idea that 'favourable climates could be manufactured' came to be largely accepted (Harrison, *Climates and Constitutions*, 22). However, recent research shows that a clear trend in this sense can already be detected in the seventeenth century (see Miglietti, 'Mastering the Climate', Chapter 4, and Morera, 'Marshes as Microclimates' in this volume).

37 Gerhard Jaritz and Verena Winiwarter, 'On the Perception of Nature in Renaissance Society', in *Nature and Society in Historical Context*, ed. Mikuláš Teich, Roy Porter and Bo Gustafsson (Cambridge: Cambridge University Press, 1997), 91–111.

38 Peter Coates, *Nature: Western Attitudes since Ancient Times* (Cambridge: Polity Press, 1998), 81.

39 For bodily knowledge, see, Richard White, 'Are You an Environmentalist or Do You Work for a Living?: Work and Nature', in *Uncommon Ground: Towards Reinventing Nature*, ed. William Cronon (New York: W.W. Norton & Company, 1995), 173.

40 See, e.g., *Reappraisals of the Scientific Revolution*, ed. David C. Lindberg and Robert S. Westman (Cambridge: Cambridge University Press, 1990); Andrew Wear, *Knowledge and Practice in English Medicine, 1550–1680* (Cambridge: Cambridge University Press, 2000); Martin, *Renaissance Meteorology*, all of which document the significant transformations taking place in various fields of science and medicine in the seventeenth century without subscribing acritically to conventional notions of a 'scientific' and 'medical' revolution but rather emphasising both continuities and discontinuities with earlier periods.

41 On the 'long decline' of astrology and humouralism, see respectively: Bernard Capp, *Astrology and the Popular Press: English Almanacs, 1500–1800* (London: Faber and Faber, 1979) and Harrison, *Climates and Constitutions*.

42 Sörlin, 'Reconfiguring Environmental Expertise', 16.

43 Miller and Rose, *Governing the Present*, 15–16.

44 Keith Lindley, *Fenland Riots and the English Revolution* (London: Heinmann, 1982); Clive Holmes, 'Drainers and Fenmen: The Problem of Popular Political Consciousness in the Seventeenth Century', in *Order and Disorder in Early Modern England*, ed. Anthony Fletcher and John Stevenson (Cambridge: Cambridge University Press, 1985), 166–195; H.C. Darby, *The Changing Fenland* (Cambridge: Cambridge University Press, 1983).

45 Nicholas Blomley, 'Making Private Property: Enclosure, Common Right and the Work of Hedges', *Rural History* 18, no. 1 (2007), 1–21.

46 Ellen F. Arnold, *Negotiating the Landscape: Environment and Monastic Identity in the Medieval Ardennes* (Philadelphia: University of Pennsylvania Press, 2012); Stephen Rippon, '"Uncommonly Rich and Fertile" or "Not Very Salubrious"? The Perception and Value of Wetland Landscapes', *Landscapes* 1 (2009): 50; Vittoria Di Palma, *Wasteland: A History* (New Haven, CT: Yale University Press, 2014), 89–92.

47 Richard Hoffmann, *An Environmental History of Medieval Europe* (Cambridge: Cambridge University Press, 2013), 138.

48 Carole Rawcliffe, *Urban Bodies: Communal Health in Late Medieval English Towns and Cities* (Woodbridge: The Boydell Press, 2013); Leona Skelton, *Sanitation in Urban Britain, 1560–1700* (London: Routledge, 2016).

49 Keith Thomas, *Man and the Natural World: Changing Attitudes in England, 1500–1800* (London: Penguin, 1983), 254–255.

50 David Garrioch, '1666 and London's Fire History: A Re-Evaluation', *The Historical Journal* 59, no. 2 (2016), 319–338.

51 Paul Warde, 'Fear of Wood Shortage and the Reality of the Woodland in Europe, c. 1450–1850', *History Workshop Journal* 62 (2006), 28–57. Standish cited in Warde, 'The Invention of Sustainability', 160.

52 Carolyn Merchant, *Reinventing Eden: The Fate of Nature in Western Culture* (New York: Routledge, 2003), 59–63.

53 Cressy Dymock, 'A Discovery for New Divisions, or, Setting out of Lands, as to the Best Forme: Imparted in a Letter to Samuel Hartlib, Esquire', in *A Discoverie for Division or Setting Out of Land, as to the Best form Published by Samuel Hartlib Esquire, for Direction and*

More Advantage and Profit of the Adventurers and Planters in the Fens and Other Waste and Undisposed Places in England and Ireland, ed. Samuel Hartlib (London: Richard Wodenothe, 1653), 1–11.

54 Richard Grove, 'Cressey Dymock and the Draining of the Fens: An Early Agricultural Model', *The Geographical Journal* 147, no. 1 (1981), 27–37.

55 Paul Slack, *The Invention of Improvement: Information and Material Progress in Seventeenth-Century England* (Oxford: Oxford University Press, 2014).

56 Torsten Meyer, 'Cultivating the Landscape: The Perception and Description of Work in Sixteenth- to Eighteenth-Century German "Household Literature" (*Hausväterliteratur*)', in *The Idea of Work in Europe from Antiquity to Modern Times*, ed. Josef Ehmer and Catharina Lis (Farnham: Ashgate, 2009), 222.

57 Walter Blith, *The English Improver, or a New Survey of Husbandry* (London: J. Wright, 1649), Sig. A1v.

58 43 Elizabeth I, c. 11, 'An Act for the Recovery of Many Hundred Thousand Acres of Marshes and other Groundes', in *Statutes of the Realm*, 11 vols. (London: HMSO, 1810), iv, 977; 'May 1649: An Act for Drayning the Great Level of the Fens, Extending Itself into the Counties of Northampton, Norfolk, Suffolk, Lincoln, Cambridge and Huntingdon, and the Isle of Ely, or Some of Them', in *Acts and Ordinances of the Interregnum, 1642–1660*, ed. C.H. Firth and R.S. Rait (London: HMSO, 1911), 130–139.

59 Timothy Nourse, *Campania Foelix, or, a Discourse of the Benefits and Improvements of Husbandry Containing Directions for All Manner of Tillage, Pasturage, and Plantation* (London: Thomas Bennet, 1700), 4.

60 Glacken, *Traces on the Rhodian Shore*, 359.

61 Vera Candiani, *Dreaming of Dry Land: Environmental Transformation in Colonial Mexico City* (Stanford: Stanford University Press, 2014); Simon Pooley, *Burning Table Mountain: An Environmental History of Fire on the Cape Peninsula* (Basingstoke: Palgrave Macmillan, 2014), 16–28.

62 Candiani, *Dreaming of Dry Land*, 31; Alan Mikhail, *Nature and Empire in Ottoman Egypt: An Environmental History* (Cambridge: Cambridge University Press, 2013), 39.

63 Grove, *Green Imperialism*.

64 Ibid.

65 Mikhail, *Nature and Empire in Ottoman Egypt*; White, *The Climate of Rebellion*.

66 Wenkai He, 'Public Interest and the Financing of Local Water Control in Qing China, 1750–1850', *Social Science History* 39, no. 3 (2015), 409–430.

67 Lucien Febvre with Lionel Bataillon, *A Geographical Introduction to History* (London: Kegan Paul, Trench Trubner & Co., 1925), 1–17.

68 Glacken, *Traces on the Rhodian Shore*, 565–566.

69 Philip Slavin, 'The Great Bovine Pestilence and Its Economic and Environmental Consequences in England and Wales, 1318–50', *Economic History Review* 65, no. 4 (2012), 1239 1266; Dolly Jørgensen, 'Running Amuck? Urban Swine Management in Late Medieval England', *Agricultural History* 87, no. 4 (2013), 429–451; Alan R. Macdonald and John McCallum, 'The Evidence for Early Seventeenth-Century Climate from Scottish Ecclesiastical Records', *Environment and History* 19, no. 4 (2013), 487–509.

70 Exceptions include Angus Winchester, *The Harvest of the Hills: Rural Life in Northern England and the Scottish Borders, 1400–1700* (Edinburgh: Edinburgh University Press, 2000).

71 Donald Worster, 'Doing Environmental History', in *The Ends of the Earth: Perspectives on Modern Environmental History*, ed. Donald Worster (Cambridge: Cambridge University Press, 1988), 289–307.

72 See, e.g., Eric T. Jennings, *Curing the Colonizers: Hydrotherapy, Climatology, and French Colonial Spas* (Purdue: Duke University Press, 2006); Janković, *Confronting the Climate*; Fressoz, *L'Apocalypse joyeuse*.

73 On shifting attitudes towards the human body, see Harrison, *Climates and Constitutions*.

1 Climate, travel and colonialism in the early modern world

Rebecca Earle

Introduction: travelling bodies

In an oft-reprinted early modern text, the Spanish humanist Diego Rodríguez de Almela offered the following thoughts on the emotional ties that bind men to their patria, or homeland:

> The love that men feel for the land where they were born or raised forms part of their very nature . . . and wise men even say that there are certain ailments that can afflict men far from the land where they were born and raised that can be cured only by returning to that land. This is because their complexions suit the air of the place where they were raised and different airs can and do make men ill; and this affects even the dead, for they say that cadavers rest more easily in the lands where their forefathers are buried than in any other.[1]

Many writers in the early modern era made similar comments about the deep love that individuals naturally felt for their homeland.[2] The author of a late-sixteenth-century general history of mankind thus included a chapter entitled 'How all men feel deep love and longing for their natal soil in which they were born and raised'. Diego Rodríguez de Almela's insistence that men who absented themselves from this soil risked all manner of illness was widely shared.[3]

This, of course, was precisely the era in which Europeans were embarking on overseas travel on an unprecedented scale, which took them first to Africa and then to the Americas and Asia. The Portuguese had been exploring the west coast of Africa since the early fifteenth century, and in 1488 Portuguese navigators rounded the Cape of Good Hope to reach the Indian Ocean. In 1492 Columbus's crew sailed in a south-westerly direction hoping to reach Asia; a few years later Vasco de Gama's fleet reached India by sailing east. Between 1519 and 1522 the crew of Ferdinand Magellan's fleet of five ships circumnavigated the globe, and for the next century European sailors and explorers ventured ever farther from their homelands in pursuit of trade and conquest. It is thus a notable feature of Europe's colonial expansion in the fifteenth and sixteenth centuries that it occurred in a moment in which European writers were placing particular emphasis on the deep love that individuals naturally felt for their natal soil. The voyages of discovery and subsequent colonial ventures, in other words, took place at a time when conventional wisdom affirmed that prolonged absence from home was likely to induce almost unbearable homesickness. Of course great things could not be expected from those timorous souls who remained forever at home, but venturing overseas was nonetheless recognized as an emotionally wrenching and therefore dangerous experience.

How did Europeans negotiate these challenges during the age of discovery? This chapter charts the ways in which settlers sought to mitigate against these threats to health and corporeal integrity, using the Spanish experience in the new world to illustrate broader attitudes characteristic of travellers and colonists from many parts of western Europe. These ideas about the dangers facing the body in motion form part of the history of early modern European expansion. They also shed light on the ideologies that underpinned the colonialisms that followed in its wake. Early modern Europeans perceived many similarities between the ways their bodies responded to displacement and the responses of other bodies set in motion by European colonisation and the Atlantic slave trade. Travelling bodies therefore lie at the heart of the relationship between colonialism and the emergence of race.

Scholars have long recognised the threat that unfamiliar climates were believed to pose to the European body and have increasingly drawn connections between European anxieties over the detrimental impact of colonial environments and the emergence of racial ideologies. Building on the foundational research of Antontello Gerbi and Clarence Glacken, many scholars have studied how European critiques of 'tropical' environments resonated with critiques of the peoples who inhabited these regions.[4] A significant and stimulating body of research now asserts that the idea of race grew out of this dual critique of extra-European environments and extra-European peoples. In a series of influential works, Jorge Cañizares Esguerra for instance argued that colonists in early modern Spanish America came to believe that their bodies differed in permanent and fundamental ways from the bodies of Amerindians. In his view, the driving force behind this early articulation of 'race' was the American climate. Cañizares Esguerra maintained that colonial scholars posited a radical discontinuity between European and indigenous bodies because they could not otherwise account for the fact that Europeans appeared to thrive in the, American environment, while Amerindians sickened and died.[5] In his analysis of colonial Peru Bernard Lavallé likewise argued that in the view of settlers, 'the American climate' was responsible for the corporeal differences that distinguished Europeans from native peoples.[6] Scholars working on Anglo-America have advanced similar arguments about the seventeenth-century English settlements. The scholarship on eighteenth-century 'climatic determinism' similarly posits close links between the intertwined histories of climate, colonies and race.[7] Such interpretations have provoked a lively debate about the origins of racial thinking and its possible links to early modern colonial expansion. Underpinning much of this debate is the conviction that the idea of race is closely associated with European responses to the extra-European environment.

This research highlights the dilemmas that overseas colonisation posed to Europeans and helpfully focuses attention on the fact that early colonial actors ascribed great significance to the differences they perceived between their bodies and those of peoples they encountered as they voyaged ever farther from their own lands. Nonetheless, an unfamiliar climate was but one of the challenges faced by the early modern traveller. By separating climate from the larger humoral framework that animated all early modern understandings of corporeality, we risk losing an appreciation of the underlying ideas that helped Europeans make sense of these differences. Shifting from a singular focus on climate permits a fuller appreciation of the ways that early modern Europeans believed that travelling bodies of all sorts interacted with their environments.

'The safety and health of travailers'[8]

Leaving one's natal soil was dangerous because, as Rodríguez de Almela explained, each person's complexion was best suited to the air, water and food of their homeland. Humoral theory, which provided an understanding of the human body universally embraced in early modern Europe, offered a coherent explanation of why this should be the case. The origins of humoralism lay in the writings of classical authors such as Hippocrates, but, as the Introduction argues, it offered an enduring model for making sense of the body's interactions with its environment that persisted well into the eighteenth century.[9] The humoral body was understood to consist of a balance of the four humours that both governed health and gave each person their individual complexion. Each person was born with a particular complexion or temperament, terms which referred equally to both physical appearance and character, but the particular balance of humours acquired at birth was unlikely to remain constant throughout one's life. Alterations in an individual's pattern of eating, sleeping, exercise and digestion, in the airs and waters that surrounded them, or in their emotional equanimity could induce imbalances in their humours, which could in turn provoke changes in mood, well-being and character. The complexion was thus changeable, varying both over the course of a lifetime and in accordance with changes in lifestyle or environment.

Such transformations, however, were fraught with danger. Sudden changes of any sort were to be avoided. 'Changing your habits can be lethal', ran the saying.[10] Doctors liked to cite Hippocrates' warning that even healthy people could be harmed by an abrupt alteration and that any shift, even from a bad to a good diet, was potentially dangerous.[11] Only with great care should an individual alter their basic complexion by introducing changes into their regimen, thereby acquiring a 'second nature'. Indeed, for this reason some writers argued that it was best to ensure that one's normal diet was not too limited, as otherwise the slightest disturbance in the availability of food could prove dangerous. It was better slowly to accustom oneself to a variety of foods rather than be reliant on only a handful of foodstuffs.[12]

Changes in environment were equally challenging to the humoral body. Since the time of Hippocrates European writers had drawn connections between the environment in which individuals lived and their characters, and during the sixteenth and seventeenth centuries the influence of climate on the human constitution was universally acknowledged. As one Spanish scholar put it in 1608, 'people to a certain extent resemble the place where they are born'.[13] This meant that in general people were ideally suited to their home environment. At the very least, exposure to an unfamiliar climate was likely to upset the balance of humours, thereby causing illness. Some climates were inherently more healthy than others – damp, swampy places were generally viewed as dangerous – but it was considered unwise to undergo sudden alterations of environment, even from an unhealthy to a more salubrious climate, just as it was dangerous to alter one's diet precipitously. More dramatically, prolonged residence in a different environment might provoke significant transformations of the overall complexion. Writing in the early seventeenth century, the Dominican priest Gregorio García explained that although Ethiopians were, like all men, the sons of Noah (who García believed had been white), because they now lived in the heat of the torrid zone, their skin had darkened. Lengthy exposure to a hot climate profoundly altered their appearance.[14] Overall, in the words of a German cosmographer, a change in climate could result in a change in 'talent, vivacity and condition'.[15]

Clearly, long-distance travel was likely to present the traveller with both new climates and unfamiliar foods. As one seventeenth-century writer warned, such changes were challenging even to the most robust complexions. Travel was particularly ill advised for old men, who, he insisted, 'shall never be able to endure the frequent changes of diet and aire, which young men cannot bear without prejudice to their health, except it be little and little and (as it were) by insensible degrees'.[16] Abrupt changes in eating patterns were perhaps the most obvious challenge to the traveller, and for this reason medical writers had long advised that travellers take particular care over their diets. For example, the health manual composed by the ninth-century Baghdad-based physician Qusta ibn Luqa for a client intending to undertake a pilgrimage to Mecca offered detailed advice on which foods were most suitable in helping the body cope with the rigours of travel. He stressed that keeping to a healthy diet was vital if the traveller was to avoid illness. An appropriate diet, in turn, would protect against sickness. 'If he keeps to this regimen the humours of his body will not become sharp and no fatigue or other diseases will befall him, which originate from the intense movement during a long journey', the doctor noted.[17] Ideally, in fact, the traveller would bring his own supply of food, precisely to ensure that he did not further stress his already fatigued body with strange foodstuffs.

Early modern advice echoed this view. Such travel guides repeated cautionary tales drawn from both classical antiquity and more recent history to illustrate the dangers that befell travellers who failed to attend sufficiently to their regimen. For instance, sudden changes of clothing were extremely ill advised. One guide to foreign travel warned of the case of 'One *Tho. Randolph* in Queen *Elizabeths* time being Embasssador to *Ivan Vasilonoch* Emperor of *Moscovia*, reporteth himselfe to have narrowly avoided death or extreame sicknesse, which he had gotten, by changing his English habit over-quickly into a Russian apparel'.[18] As Thomas Neale, the guide's author, explained, 'the sodaine change of any habituated Custome is so dangerous, that sometimes ordinary things disused bring on the necessity of death'. Neale therefore advised that travellers adapt themselves slowly and 'by degrees' to the customs of their destination. In this way, he assured his readers, they could avoid illness or, indeed, death. Avoiding 'over-quick' alterations to one's usual routine was thus essential for safe travel. Writing in the early seventeenth century, the English aristocrat Thomas Palmer stipulated that during travel, 'let the diet of all men, for eating, drinking, sleeping, clothing and such like, be answerable to every ones nature'. By this means travellers could ensure that they 'always keep themselves in one temper'. Summing up what was by then long-standing wisdom, he overall advised that travellers 'therefore observe three preventers of mischiefs, and inconveniences to [their] safety and health . . . namely, Diet . . . Exercise, and moderation of Passions'.[19]

Unaccustomed clothing, new environments and disruptions to daily routine were not the only threats facing the early modern traveller. As writers such as Diego Rodríguez de Almela indicated, the sadness and pain caused by being far from one's homeland in themselves posed a serious danger to well-being. This, indeed, was why Palmer stressed the need to maintain an emotional equilibrium when travelling. Sadness had long been recognized as a significant threat to health; a rich corpus of texts discussed the effects of melancholy on the body and considered ways of managing its impact.[20] Strikingly, early modern writings often noted that absence from one's homeland was liable to induce sadness and melancholia. In contrast to earlier truths that had affirmed, in the words of the ancient playwright Menander, '*ubi bene ibi patria*' [home is wherever one is happy], early modern thinkers increasingly insisted that travel away from one's homeland was a deeply unpleasant experience.[21] The author of one late-sixteenth-century history noted that there was no need to elaborate on this fact, as it was 'perfectly evident, something that we all see daily and indeed hourly in those who are absent from their own homeland'.[22]

Homesickness, together with the changes to regimen usually provoked by leaving home, made early modern travel a particularly dangerous undertaking. As the Spanish jurist Juan de Solórzano y Pereira observed, experience provided ample evidence of the 'illness, harm and death that usually result from leaving the climate and place where we were born and raised'. Love of homeland, he continued, was so powerful that often the only cure was for the patient to return home to breathe its restorative air.[23] Writers in other words viewed travel as dangerous both because of the changes it imposed on an individual's daily routine and also because of the sadness that was assumed to afflict anyone far from their natal soil, which was itself liable to induce illness.[24] Clearly, the delicate European body required protection from the rigours of travel.

'The usual foods we eat in Spain'

The principle bulwark on which settlers and explorers relied to shield themselves from travel-induced illness was diet. Given the humoral understanding of the human body that underpinned early modern epistemologies, this made perfect sense. The perturbations provoked by unfamiliar airs and waters could best be offset by maintaining consistency in diet because of food's central role in maintaining the overall complexion. From the earliest days of Spanish overseas expansion colonists constantly asserted that European food was an essential defence against illness and early death. In 1493, Columbus had insisted that his settlers would die were they not provided with 'the usual foods we eat in Spain', and countless subsequent colonists echoed his sentiments.[25] For this reason the familiar foods of Europe took on a positively totemic importance in colonial writings as guarantors of health and well-being. Spanish writers focused particularly on the structural elements of the ideal Iberian diet: wine, wheat bread, olive oil and meat. 'To deprive an old man or a youth of a little wine', observed one official in sixteenth-century Guatemala, 'is to send him straight to the grave'.[26] Wheat bread, which also served as a crucial symbol of Catholic identity through its role in the Mass, was proclaimed an absolute necessity, and Spanish settlers regularly denounced new-world carbohydrates such as maize or cassava as utterly unsuited to their constitution. Settlers in Florida for example insisted that sick people in particular 'cannot under any circumstances eat the said maize, and a number of people have died because they had no other food'.[27] As the seventeenth-century Jesuit writer Bernabé Cobo noted, these foods were simply not 'suitable to sustain Spaniards'.[28] What European settlers needed was the familiar, healthful foods of home.

Settlers therefore went to considerable lengths to obtain these health-giving foods. From the earliest decades of the colony the ships that travelled to the Indies from Spain were laden with the red wine, olive oil, wheat flour and other foodstuffs necessary for the reproduction of the Iberian diet, and royal orders repeatedly exhorted settlers to plant wheat and other key crops. Indeed, only those regions in which European plants grew well were considered suitable for Spanish colonisation.[29] Grants of land and indigenous labour specifically stipulated that their holders must cultivate European crops, and conquistadors were required to provide their troops with appropriate European foodstuffs. Spanish settlers indeed planted wheat, radishes, barley, cabbage and a host of other old-world plants up and down the Americas and were equally energetic in introducing old-world livestock such as cattle, pigs and sheep to both the Caribbean islands and the American mainland.[30]

These familiar foods were doubly medicinal, for in addition to being well suited to the Spanish body they also helped remedy the sadness that men far from home were likely to experience. As the chronicler Antonio de Herrera noted, settlers fell ill not only from the 'change to such a different climate' and 'the local food' but also because they were saddened 'to find themselves so far from their own lands'.[31] Melancholy expatriates were thus in particular need of familiar, sustaining foods. For the same reason slave traders, concerned for the health of their valuable human property, made efforts to supply familiar food such as plantains, rice, yams and couscous to slaves, particularly those who were already weakened by illness. Such measures were important, as melancholy sadness was acknowledged to be a condition to which enslaved people, not surprisingly, were prone.[32] In other words, travelling Spaniards and enslaved Africans alike were liable to fall victim to the dangerous effects of homesickness and could alike benefit from restorative familiar foods.

In the case of Spanish settlers, it was not simply that these foodstuffs helped maintain individual health. Such foods helped preserve their specifically *Spanish* complexion despite their distance from the Peninsula. Food helped make Spaniards Spanish in several ways. To begin with, it was of course important in forging the affective bonds that linked a man to his homeland. Writers rhapsodised about the love individuals naturally felt for the place 'where their body gained the strength to take its first steps, whose air formed their first breath, *where they ate their first meals*' and where they spent their childhood.[33] This was in part why eating familiar foods could help alleviate homesickness. Food's significance to the construction of a specifically national identity, however, transcended this general affective relationship between person and place. As early modern writers insisted, diet, together with the climate and other aspects of an individual's lifestyle, not only explained the particular contours of the individual complexion but also determined the ways in which, say, Spaniards, as a group, differed from the inhabitants of other states and kingdoms. Drawing on standard Galenic principles, the physician Juan Huarte de San Juan for example explained that men differed one from another 'by reason of the heat, the coldness, the moisture, and the drouth, of the territorie where men inhabit, of the meats which they feed on, of the waters which they drink, and of the aire which they breath'. He then observed that these factors determined the ways in which Spaniards differed from Frenchmen.[34] As long as they continued to consume Spanish food, in other words, colonists would remain essentially Spanish.

Settlers who persisted in building a diet around new-world foods despite such recommendations risked even graver perturbations in their overall complexion. There was widespread agreement that Europeans who relied on indigenous staples such as maize risked going native not only in their customs but also in their very bodies. The occasional pineapple or avocado was considered harmless enough; indeed, colonists waxed lyrical about the delights of such tropical fruits. Starchy alternatives to European grains were another matter. Breads and porridges made from wheat, barley and oats formed the core of the European diet and therefore underpinned the European constitution. Altering one's diet by incorporating a little chile pepper did not pose the same risks as substituting maize or cassava for these familiar and appropriate staples. Warnings about the dangerous effect of new-world foods therefore focused on the 'breads of the indies', as colonial texts routinely dubbed maize, cassava and other new-world starches. Writer after writer insisted that the European constitution simply could not thrive on these novel foods.[35]

Settlers saw evidence for their dangerous effects all around them. 'Here we have seen very white men from Spain', warned one writer in seventeenth-century Peru, 'who, on withdrawing into the hills and eating maize and other Indian dainties, return so toasted that they resemble Indians'. This author, like many others, believed that Amerindians had originated in the old world and that their transformation into a 'toasted' people was the direct result of a diet of maize and cassava, together with their exposure to the American environment.[36] New foods, as much as new climates, posed a fundamental threat to the European body. Familiar foods provided the surest means of defence.

The consequences of a careless deviation from one's usual regime were made clear not only in medical but also in legal texts. The Italian jurist Baldus de Ubaldis, author of a number of influential legal commentaries, for instance treated citizenship as a form of individual complexion. He argued that people born in a particular place naturally developed a *habitus* or habitual inclination toward that place, which was an important component of citizenship. Residence in a different location resulted in the development of a 'second nature' that transformed an individual's habitus and allowed them to become citizens of their new domicile. 'Second nature' is precisely the phrase used in humoral medicine to describe the transformations worked on the complexion by the consumption of unfamiliar foods, travel to different climates, or adoption of a radically different lifestyle. Civic or patriotic identity was thus at least in part an embodied condition that might be altered by the experience of living elsewhere. Baldus explicitly compared the acquisition of a new civic identity to the changes that occurred in a plant moved from its original habitat into a new setting. Residence in a new location could thus induce alterations in many aspects of the individual temperament, from physical form to emotional attachment.[37]

In summary, food was understood to be a key element in maintaining the health of colonial populations, because familiar foods best suited each person's constitution, because they helped counteract the threats posed by homesickness and also because they helped preserve the European temperament even at considerable distance from Europe. Little wonder that colonists up and down the Indies and beyond worked hard to 'Europeanize' colonial landscapes, to use Alfred Crosby's term, by introducing European agriculture, and that the ships that sailed from Spain to Havana and Veracruz were laden with red wine, olive oil, wheat flour and the other staples of Iberian cuisine.[38] These foods were the front line in the defence against degeneration and death among the settler class.

They 'succumbed to the change of climate, country, and food'

Unsurprisingly, Europeans believe that travel had the same detrimental effect on Amerindians that they observed in their own bodies and in those of enslaved Africans. Abrupt movement from one climate to another, together with changes in diet, were frequently blamed for illness among colonised populations in the Americas. The humanist scholar Peter Martyr for example noted that of the ten indigenous interpreters taken from the Caribbean to Spain after Columbus's second voyage, 'only three survived; the others having succumbed to the change of climate, country, and food'.[39] The Dominican priest Reginaldo de Lizárraga likewise observed that when residents of the high Andes were forcibly relocated to the lowlands they sickened, 'as occurs everywhere'.[40] One hundred fifty years later, Caribbean planters continued to warn that, as 'has often been observed', moving enslaved labourers even from one plantation to a neighbouring one was likely to make them 'droop and be sickly'.[41] Travel, in other words, was just as dangerous for Amerindians and Africans as it was for Europeans. Indeed, given Amerindians' allegedly feeble constitution, it was more dangerous, in the view of Spaniards. Consequently the Spanish crown in 1543 prohibited colonists from transporting Amerindians to Europe, even if the Amerindians were travelling voluntarily. This decision was necessary because, as the legislation explained, 'the majority of those Indians die, because these parts are different from their own kingdoms, and contrary to their nature, and because they are of feeble complexion'.[42] The factors offered to explain this mortality were thus those offered to explain ill health in any traveller: the disruptive impact of an unfamiliar environment, with its attendant changes in diet and air.

As the 1543 legislation indicated, Spaniards tended to view Amerindians as 'feeble'. In the words of another sixteenth-century writer, they were 'delicate and feminine and of feeble complexion'. This, he explained, was due to their diet:

> They do not have the habit of eating meat, as there was none in that land aside from some little animals like rabbits, which were not enough for everyone anyway, and some parrots, and as a result they ate fishes and worms that grow in the earth, and they didn't have any wine and for this reason they died so young.[43]

This writer was not unusual in his interest in explaining mortality among Amerindians. After 1492 the indigenous population of the Americas suffered a precipitous decline. Scholarly estimates of the scale of the demographic collapse vary, but there is no doubt that in many areas it was catastrophic.[44] Spaniards were troubled by this, if for no other reason than that it reduced the size of the labour force, and they expended much effort in accounting for it. Strikingly, writers often insisted that mortality was due not to abuse by the Spanish nor to the inadequacies of the pre-conquest diet, as the doctor quoted earlier maintained, but rather to the disruptive effects of the adoption of the *Spanish* diet. The Jesuit writer José de Acosta summed up the current orthodoxy when he observed in the 1590s that 'people attribute [the decline in the indigenous population] to various causes, some to the fact that the Indians have been overworked, *others to the changes of food and drink that they adopted after becoming accustomed to Spanish habits*, and others to the excessive vice that they display in drink and other abuses'.[45] New foods, in other words, were just as dangerous for indigenous bodies as they were for Europeans. If the adoption of new foods was combined with travel the effect could be lethal.

The Spanish doctor Juan de Cárdenas offered similar explanations as to why the Chichimec Indians, who in their own environment were hardy and robust, fell ill and died when incorporated into colonial society. He attributed their mortality to various causes, first among them 'the change in food, in that they are deprived of the natural sustenance on which they were raised, which, although it is very bad in itself, is for them healthy and very good, as they are accustomed to it, unlike our food which harms them'. Dreadful though the Chichimec diet was (Cárdenas explained that it consisted largely of raw meat), it was better suited to their complexion than was European food. 'As our food is foreign and harmful to them, it does not give them strength to resist illness', he concluded.[46]

In keeping with basic humoral principles, Cárdenas also observed that changes in their level of exercise and their unhappy emotional state following incorporation into colonial society exacerbated their ill health. He noted in particular the bad effects produced by 'the sad rage and melancholy that overcomes them, on finding themselves among men whom they loath so much'. As doctors since Galen had warned, sadness and melancholy were liable to induce all sorts of illness. Other colonial writers agreed that relocated Amerindians tended to sicken, 'as they miss the nature of the climate in which they were born and raised, and despair at having to leave the lands that they used to cultivate'.[47] Amerindians, just like Europeans, thus fell victim to homesickness and the potentially fatal sadness it induced. Travel in the early modern era was thus a dangerous undertaking regardless of whether the travelling body was European, African or indigenous.

In this regard, the bodies of Amerindians and West Africans were not posited by early modern European writers as essentially different from their own. Early modern colonisation did not rely on ideas about the fundamental incommensurability of European and non-European bodies but rather on an epistemological framework that stressed their underlying similarities. Non-European bodies perhaps differed from European bodies, but these differences were the result of fundamental similarities, which rendered all bodies subject to the imperious reign of the temperament and the humours. In the end, such bodies were more similar than they were different.

Conclusions: because of the climate or the air or the food . . .

European understandings of how the human body operated thus influenced how Europeans made sense of the experience of overseas colonisation. Because humoralism played such a central role in how all Europeans made sense of their bodies, concerns about the corporeal effects of travel, and in particular of new foods, were widely expressed not only by Iberian colonists in the Indies but also by many other categories of traveller. As one English writer insisted, due to the hot climate, 'an *European* can hardly live in *Aetheopia* or under the Equinoctiall line above five years'.[48] Europeans living in the hot climate of equatorial Africa thus imported quantities of wheat flour and wine in order to protect their health. 'If the ships which bring these goods did not come, the white merchants would die, because they are not accustomed to negro food', explained a Portuguese pilot who left a short account of the nascent sugar industry in São Tomé.[49] Travel in Europe itself posed significant dangers even to European travellers. Writers warned that English pilgrims travelling to Santiago de Compostella risked falling ill from 'eatynge of frutes and drynkynge of water', and English troops campaigning in northern Spain complained that the region's cider made them ill.[50] Nor were Spaniards the only ones to worry that through travel they risked losing their very identity. 'In this point let experience be consulted with; her unpartiall sentence shall easily tell us, how few young travellers have brought home, sound and strong, and (in a word) *English* bodies', warned the English cleric Joseph Hall in 1617.[51]

Moreover the perceived fragility of the European body in the age of discovery should encourage us to reconsider the idea that European colonisers were clad in the invincible armour of impenetrable self-confidence. Early modern colonialism was in many ways an anxious pursuit, not least because the health and stability of the European body was always in doubt. As colonists noted, the combination of unusual foods and an unfamiliar climate claimed the lives of many settlers; 'everyone from Spain is struck with a *chapetonada*, which kills more than a third of the people who come here', observed one settler in 1570s Mexico, although this did not deter him from urging his relatives to follow him to the new world.[52] Those who did not die risked equally disturbing transformations. 'Whether because of the climate or its air, or because of its foods, those who live [in the Americas] become like their surroundings, and even worse: liars, swindlers, cheats, traitors, ambitious, proud men who seek power by any means, no matter how illicit', insisted the jurist Baltasar Álamos de Barrientos at the end of the sixteenth century.[53] Colonial writers debated whether such changes were inevitable, and many argued that the new world's climate was no less healthful than that of Europe, but merely by being different it constituted a threat to the European body.

Europeans thus believed that their bodies responded to travel and the encounter with new climates in the same ways that the bodies of other peoples did, and they attempted to mitigate against these challenges in the same ways that they advised others to do: by attending to their daily regimen and especially to their diet. The underlying similarities linking all bodies reflected a robust and long-lasting understanding of the human constitution as composed of a balance of humours and in constant dialogue with its larger environment. Early modern colonialisms and the profoundly unequal societies that they constructed were not premised on the idea that different populations possessed fundamentally incommensurable bodies but rather that all bodies, however different, shared certain common features.

Acknowledgement

Chapter 1, 'Climate, travel and colonialism in the early modern world', partly reproduced with permission of Palgrave Macmillan from Rebecca Earle, 'Diet, travel and colonialism in the early modern world', in *Global Goods and the Spanish Empire, 1492–1824: Circulation, Resistance and Diversity*, ed. Bethany Aram and Bartolomé Yun-Casalilla (Basingstoke: Palgrave Macmillan, 2014), pp. 137–52.

Notes

1 Diego Rodríguez de Almela, *Valerio de las historias escolásticas*, ed. Fernan Pérez de Guzman (Salamanca, n.p., 1587 [1462]), 159.

2 See in particular José Antonio Maravall, *Estado moderno y mentalidad social, siglos XV a XVII*, 2 vols. (Madrid: Ediciones de la Revista de Occidente, 1972), I:468–476.

3 Juan Sánchez, *Corónica y historia general del hombre* (Madrid, 1598), 47–48 (quote). As William Vaughan noted, 'that which is a man's native soyle and countries aire is best': William Vaughan, *Approved Directions for Health* (London, n.p., 1612), 2. Or see Thomas Neale, *A Treatise of Direction, How to Travell Safely and Profitably into Forraigne Countries* (London, n.p., 1643), 18.

4 Antonello Gerbi, *The Dispute of the New World: The History of a Polemic, 1750–1900*, trans. Jeremy Moyle (Pittsburgh: University of Pittsburgh Press, 1973 [1955]); Clarence Glacken, *Traces on the Rhodian Shore: Nature and Culture in Western Thought from Ancient Times to the End of the Eighteenth Century* (Berkeley: University of California Press, 1967); and, for a review of scholarship on the 'tropics', David Arnold, *The Problem of Nature: Environment, Culture and European Expansion* (Oxford: Blackwell, 1996).

5 Jorge Cañizares Esguerra, 'New Worlds, New Stars: Patriotic Astrology and the Invention of Indian and Creole Bodies in Colonial Spanish America, 1600–1650', *American Historical Review* 104, no. 1 (1999): 33–68; Jorge Cañizares Esguerra, *Nature, Empire, and Nation: Explorations of the History of Science in the Iberian World* (Stanford: Stanford University Press, 2006); and Jorge Cañizares Esguerra, 'Demons, Stars and the Imagination: The Early Modern Body in the Tropics', *The Origins of Racism in the West*, ed. Miriam Eliav-Feldon, Benjamin Isaac and Joseph Ziegler (Cambridge: Cambridge University Press, 2009), 313–325.

6 Bernard Lavallé, *Las promesas ambiguas. Criollismo colonial en los Andes* (Lima: Pontífica Universidad Católica del Perú, 1993), 53 (quote), 57; or Solange Alberro, *De gachupín al criollo, O cómo los españoles de México dejaron de serlo* (Mexico City: El Colegio de México, 1992), 46. For debate over the 'Hispanic origins of race' see also Carlos López Beltrán, 'Hippocratic Bodies: Temperament and Castas in Spanish America (1570–1820)', *Journal of Spanish Cultural Studies* 8, no. 2 (2007), 253–289; Irene Silverblatt, 'Foreword', in *Imperial Subjects: Race and Identity in Colonial Latin America*, ed. Andrew Fisher and Matthew O'Hara (Durham: Duke University Press, 2009), ix–xii; Ruth Hill, 'Entering and Exiting Blackness: A Color Controversy in Eighteenth-Century Spain', *Journal of Spanish Cultural Studies* 10, no. 1 (2009): 43–58; Ralph Bauer and José Antonio Mazzotti, 'Introduction', in *Creole Subjects in the Colonial Americas*, ed. Ralph Bauer and José Antonio Mazzotti (Chapel Hill: University of North Carolina Press, 2009), 1–58; Max S. Hering Torres, María Elena Martínez and David Nirenberg, eds., *Race and Blood in the Iberian World* (Vienna: LIT Verlag, 2012); Rebecca Earle, *The Body of the Conquistador: Food, Race and the Colonial Experience in Spanish America, 1492–1700* (Cambridge: Cambridge University Press, 2012); and Kimberly Anne Coles, Ralph Bauer, Zita Nunes and Carla L. Peterson, eds., *The Cultural Politics of Blood, 1500–1900* (London: Palgrave Macmillan, 2015).

7 Karen Ordahl Kupperman, 'Fear of Hot Climates in the Anglo-American Colonial Experience', *William and Mary Quarterly* 41, no. 2 (1984): 213–240; Joyce Chaplin, 'Natural Philosophy and an Early Racial Idiom in North America: Comparing English and Indian Bodies', *William and Mary Quarterly* 54, no. 1 (1997): 229–252; and Joyce Chaplin, *Subject Matter: Technology, the Body, and Science on the Anglo-American Frontier, 1500–1676* (Cambridge: Harvard University Press, 2001). For a sampling of important works addressing the relationship between climate, race and colonialism, see also James Sweet, 'The Iberian Roots of American Racist Thought', *William and Mary Quarterly* 54, no. 1 (1997): 143–166; Mark Harrison, *Climates and Constitutions: Health, Race, Environment and British Imperialism in India, 1600–1850* (Oxford: Oxford University Press, 1999); Catherine Hall, 'Introduction: Thinking the Postcolonial, Thinking the Empire', in *Cultures of Empire: A Reader: Colonizers in Britain and the Empire in the Nineteenth and Twentieth Centuries*, ed. Catherine Hall (New

York: Taylor & Francis, 2000), 1–36; Roxanne Wheeler, *The Complexion of Race: Categories of Difference in Eighteenth-Century British Culture* (Philadelphia: University of Pennsylvania Press, 2000); Mary Floyd-Wilson, *English Ethnicity and Race in Early Modern Drama* (Cambridge: Cambridge University Press, 2003); and Margaret Greer, Walter Mignolo and Maureen Quilligan, eds., *Rereading the Black Legend: The Discourses of Religious and Racial Difference in the Renaissances Empires* (Chicago: University of Chicago Press, 2007).

8 Thomas Palmer, *An Essay on the Means How to Make our Traveiles, into Forraine Countries, the More Profitable and Honourable* (London, n.p., 1606), 46.

9 For a clear introduction to humoralism with a focus on Spain see Luis Grangel, *La medicina española renacentista* (Salamanca: Universidad de Salamanca, 1980); Luis García Ballester, *La búsqueda de la salud: sanadores y enfermos en la España medieval* (Barcelona: Península, 2001); and Carmen Peña and Fernando Girón, *La prevención de la enfermedad en la España bajo medieval* (Granada: Universidad de Granada, 2006). For the pan-European influence of humoralism, see for example Nancy Siraisi, *Medieval and Early Renaissance Medicine: An Introduction to Knowledge and Practice* (Chicago: University of Chicago Press, 1990); and Lawrence Conrad, Michael Neve, Vivian Nutton, Roy Porter and Andrew Wear, *The Western Medical Tradition, 800 BC to AD 1800* (Cambridge: Cambridge University Press, 1995); and, for examples of vernacular humoralism in the Spanish empire and elsewhere in Europe, Barbara Duden, *The Woman Beneath the Skin: A Doctor's Patients in Eighteenth-Century Germany*, trans. Thomas Dunlap (Cambridge: Harvard University Press, 1991); Mary Fissel, *Vernacular Bodies: The Politics of Reproduction in Early Modern England* (Oxford: Oxford University Press, 2004); Sherry Fields, *Pestilence and Head Colds: Encountering Illness in Colonial Mexico* (New York: Columbia University Press, 2008); and John Slater, Maríaluz López-Terrada, and José Pardo-Tomás, eds., *Medical Cultures of the Early Modern Spanish Empire* (Farnham: Ashgate, 2014).

10 'Mudar costumbre es a par de muerte': Andrés Bernáldez, *Historia de los reyes católicos D. Fernando y Doña Isabel (c. 1500)*, 2 vols. (Seville: Imprenta que fué de J.M. Geofrin, 1870), I:125; *Refranes famosísimos y provechosos* (Madrid: Gráficas Reunidos, 1923 [1509]), viii; Juan de Valdés, *Diálogo de la lengua*, ed. Angel Alcalá Galve (Madrid: Biblioteca de la Universidad de Alicante, 1997 [1535]), 264; and Juan de Cárdenas, *Problemas y secretos maravillosos de las Indias* (Madrid: Ediciones Cultura Hispánica, 1945 [1591]), 202.

11 Juan Francisco Pacheco, *Question médica nuevamente ventilada si la variedad de la comida es dañosa para la conservación de la salud* (Jaen, n.p., 1646), 15; Luis Lobera de Ávila, *Banquete de nobles caballeros compuesto por Luis Lobera de Avila (1530)* (Madrid: Ediciones Castilla, 1952 [1530]), 133–136; Pedro de Mercado, *Diálogos de philosophia natural y moral* (Granada, n.p., 1574), 76v; and Ken Albala, *Eating Right in the Renaissance* (Berkeley: University of California Press, 2002), 50. See also Hippocrates, Aphorism ii, 50, *Hippocrates*, trans. W.H.S. Jones (Cambridge: Harvard University Press, 1998), IV:121. I am grateful to Peter Pormann for this last reference.

12 Blas Alvarez Miraval, *La conservación de la salud del cuerpo y del alma* (Medina del Campo, n.p., 1597), 233r-237v; and Pacheco, *Question médica.*

13 Francisco Bermúdez de Pedraza, *Antigüedad y excelencias de Granada* (Madrid, 1608), 146 (quote); Juan Huarte de San Juan, *Examen de Ingenios, or The Examination of Mens Wits* (London, 1594), 21–22; Cárdenas, *Problemas y secretos maravillosos*, 174–175; and Diego Andrés Rocha, *El orígen de los indios*, ed. José Alcina Franch (Madrid: Historia 16, 1988 [1681]), 69. See also Hippocrates, 'On Airs, Waters and Places', *The Genuine Works of Hippocrates*, trans. Francis Adams (London: Sydenham Society, 1849); and Glacken, *Traces on the Rhodian Shore.*

14 Gregorio García, *Orígen de los indios del Nuevo Mundo*, ed. Franklin Pease (Mexico City: Fondo de Cultura Económica, 1981 [1607]), 149–150. See Floyd-Wilson, *English Ethnicity and Race*, 23–47 for early modern theories about the impact of climate on skin colour.

15 Henrico Martínez, *Reportorio de los tiempos e historia natural desta Nueva España* (Mexico City: Consejo Nacional para la Cultura y las Artes, 1991 [1606]), 275.

16 Fynes Moryson, *An Itinerary Written by Fynes Moryson Gent* (London, n.p., 1617), 2.

17 Qusta ibn Luqa, *Qusta ibn Luqa's Medical Regime for the Pilgrims to Mecca. The risala fi tadbir safar al-hajj*, ed. Gerrit Bos (Leiden: Brill, 1992), l, 27.

18 Thomas Neale, *A Treatise of Direction, How to Travell Safely and Profitably into Forraigne Countries* (London, n.p., 1643), 31–32.

19 Palmer, *An Essay*, 46–47.

20 Melancholy was a complex condition; see Raymond Klibansky, Erwin Panofsky and Fritz Saxl, *Saturn and Melancholy: Studies in the History of Natural Philosophy, Religion and Art* (London: Nelson, 1964); Floyd-Wilson, *English Ethnicity and Race*; Angus Gowland, 'The Problem of Early Modern Melancholy', *Past & Present* 191 (2006): 77–120 and Susan Mattern, *Galen and the Rhetoric of Healing* (Baltimore, MD: Johns Hopkins University Press, 2008). 'No man is entirely free of melancholy', noted one doctor: Christoval Pérez de Herrera, *Proverbios morales y consejos christianos muy provechosos para concierto y espejo de la vida* (Madrid, n.p., 1618), 142.

21 For the original Greek see *Menandri Sententiae*, ed. Siegfried Jäkel (Leipzig: Bibliotheca Scriptorum Graecorum et Romanorum Teubneriana, 1964), line 735; I am grateful to Simon Swain for this reference. For a defense of the classical view, with appropriate citations, see Robert Burton, *The Anatomy of Melancholy*, ed. Holbrook Jackson (New York: Vintage, 1977 [1621]), 173–175.

22 Sánchez, *Corónica y historia general del hombre*, 47–48.

23 Juan de Solórzano Pereira, *Política indiana*, 2 vols. (Madrid, n.p., 1736 [1647]), I:88. See also Antonio de Herrera, *Historia general de los hechos de los castellanos en las islas i tierra firme del mar oceano*, ed. Mariano Cuesta Domingo, 4 vols. (Madrid: Universidad Complutense de Madrid, 1991 [1601]), I:324.

24 Bermúdez de Pedraza, *Antigüedad y excelencias de Granada*, 146; Maravall, *Estado moderno y mentalidad social*, I:476; and James Casey, *Early Modern Spain: A Social History* (London: Routledge, 1999), 194. A man's homeland 'is always in his heart, calling out to him', wrote Alonso Rodríguez from New Granada to his brother, who lived near Toledo: Alonso Rodríguez to Juan Rodríguez, Popayán, 4 Feb. 1578, in *Cartas privadas de emigrantes a Indias, 1540–1616*, ed. Enrique Otte (Mexico City: Fondo de Cultura Económica, 1996), 55. See also Peter Martyr to Juan Bautista de Anglería, 15 May 1488, *Epistolario*, ed. José López de Tori (Madrid: Imprenta Góngora, 1953), 25; Diego Pérez to Manuel Pérez, Panamá, 10 April 1573, in *Cartas privadas de emigrantes a Indias*, ed. Otte, 248; and Maravall, *Estado moderno y mentalidad social*, I:459.

25 Cristóbal Colón. 'Memorial que para los Reyes Católicos dio el Almirante a don Antonio de Torres, 30 Jan. 1494', in *Los cuatro viajes del almirante, Los cuatro viajes del almirante y su testamento*, ed. Ignacio Anzoátegui (Madrid: Espasa, 1971), 155–168 (quote 158). For further discussion see Earle, *The Body of the Conquistador*.

26 Carta de Tomás López Medel, 25 March 1551, Archivo General de Indias, Seville [henceforth AGI], Audiencia de Guatemala, legajo 9A, R. 18, n. 77, fol. 1. For an introduction to the early modern Spanish diet see Rafael Chabrán, 'Medieval Spain', in *Regional Cuisines of Medieval Europe: A Book of Essays*, ed. Melitta Weiss Adamson (New York: Routledge, 2002), 125–152.

27 Investigation into conditions in Santa Elena, 1576, in *Colonial Records of Spanish Florida*, ed. Jeannette Thurber Connor, 2 vols. (Deland: Florida Historical Society, 1925), I:154 (quote), 158, 162, 164, 168, 170, 174, 176.

28 Bernabé Cobo, *Historia del Nuevo Mundo*, c. 1653, in *Obras*, ed. Francisco Mateos, 2 vols. (Madrid: Biblioteca de Autores Españoles, 1956), I:375. For details of colonial responses to new world foods see Earle, *The Body of the Conquistador*.

29 See for example Asiento con Lucas Vásquez de Ayllón, 12 June 1523, AGI, Indiferente General, legajo 415, L.1, fol. 36.

30 Ordenanzas inéditas de Fernando Cortés, Temixtitan, 20 Mar. 1524, in *Dissertaciones*, ed. Lucas Alamán (Mexico City: Editorial Jus, 1969), I:270; Carta del contador Rodrigo de Albornoz al emperador, 15 Dec. 1525, in *Colección de documentos para la historia de México*, ed. Joaquín García Icazbalceta (Mexico City: Antigua Librería, 1858–66), I:489–490; Cédula to the governor of Tierra Firme, Valladolid, 19 Oct. 1537, AGI, Audiencia de Panama, legajo 235, libro 6, fol. 129; Real Cédula al virrey de la Nueva España, Valladolid, 23 Aug. 1538, in *Colección de documentos para la historia de la formación social de Hispanoamérica, 1493–1810*, ed. Richard Konetzke, 3 vols. (Madrid: Consejo Superior de Investigaciones Científicas, 1953), I:186; Carmelo Viñas Mey, 'Datos para la historia económica de la colonización española', *Revista nacional de economía* 44 (1923): 60–61; José Tudela de la Orden, 'Economía', in *El legado de España a América*, ed. José Tudela de la Orden, 2 vols. (Madrid: Ediciones Pegaso, 1954), II:665, 676; Eugene Lyon, 'Spain's Sixteenth-Century North American Settlement Attempts: A Neglected Aspect', *Florida Historical Quarterly* 59, no. 1 (1981): 279; and Justo L. del Río Moreno, 'El cerdo. Historia de un elemento esencial de la cultura castellana en la conquista y colonización de América (siglo XVI)', *Anuario de Estudios Americanos* 53 (1996): 10, 17–19.

31 Herrera, *Historia General*, I:324.

32 Alonso de Sandoval, *De Instauranda aethiopum salute: El mundo de la esclavitud negra en América*, ed. Angel Valtierra (Bogotá: Biblioteca de la Presidencia de Colombia, 1956 [1627]), 107; Linda Newson and Susie Minchin, 'Diets, Food Supplies and the African Slave Trade in Early Seventeenth-Century Spanish America', *The Americas* 63, no. 4 (2007): 533–556; David Eltis, Philip Morgan and David Richardson, 'Agency and Diaspora in Atlantic History: Reassessing the African Contribution to Rice Cultivation in the Americas', *American Historical Review* 112, no. 5 (2007): 1345–1347; and Judith Carney and Richard Rosomoff, *In the Shadow of Slavery: Africa's Botanical Legacy in the Atlantic World* (Berkeley: University of California Press, 2009), 68–69. Recently captured West Africans, stated the planter James Grainger, should be provided with food that is 'as little different from what they eat at home as may be': *An Essay on the More Common West-India Diseases; and the Remedies Which That Country Itself Produces: To Which Are Added Some Hints on the Management, &c. of Negroes* (London, n.p., 1764), 8.

33 Christoval Suarez de Figueroa, *El passagero: advertencias utilíssimas a la vida humana* (Barcelona, n.p., 1618), 11–12 (my emphasis).

34 Huarte, *Examen de Ingenios*, 21–22. Or see Levinus Lemnius, *The Touchstone of Complexions*, trans. T.N. (London, n.p., 1633), 25–31.

35 For extensive discussion of colonial attitudes to new world foods see Earle, *The Body of the Conquistador.*

36 Rocha, *El orígen de los indios*, 212. On the yellow colouring supposedly acquired by Spanish settlers long resident in the Indies see also George Mariscal, 'The Figure of the *Indiano* in Early Modern Spanish Culture', *Journal of Spanish Cultural Studies* 2, no. 1 (2001): 57; and on the Spanish debate about the origin of Amerindians, Earle, *The Body of the Conquistador.*

37 Julius Kirshner, 'Between Nature and Culture: An Opinion of Baldus of Perugia on Venetian Citizenship as Second Nature', *Journal of Medieval and Renaissance Studies* 9 (1979): 193, 198; Tamar Herzog, *Defining Nations: Immigrants and Citizens in Early Modern Spain and Spanish America* (New Haven: Yale University Press, 2003), 24, 71–72; and Peña and Girón, *La prevención de la enfermedad*, 40.

38 Alfred Crosby, *Ecological Imperialism: The Biological Expansion of Europe, 900–1900* (Cambridge: Canto, 1986).

39 Peter Martyr D'Anghera, *De Orbe Novo: The Eight Decades of Peter Martyr D'Anghera*, trans. Francis Augustus MacNutt, 2 vols. (New York: G.P. Putman's Sons, 1912), decade 1, book 2.

40 Reginaldo de Lizárraga, *Descripción breve de toda la tierra del Perú, Tucumán, Río de la Plata y Chile (c. 1609)*, ed. Toribio de Ortiguera (Madrid: Biblioteca de Autores Españoles, 1968), 16. Or see Alonso de Ovalle, *Histórica relación del Reyno de Chile* (Santiago: Instituto de Literatura Chilena, 1969 [1646]), 117–118.

41 Grainger, *An Essay on the More Common West-India Diseases*, 11–12.

42 Esteban Mira Caballos, *Indios y mestizos americanos en la España del siglo*, vol. 16 (Madrid: Iberoamericana, 2000), 54, 62 (quote), 66. I am grateful to Caroline Pennock for this reference.

43 Ruy Díaz de Isla, *Tractado contra el mal serpentino que vulgarmente en España es llamado bubas* (Seville, n.p., 1539), 39–40.

44 Noble David Cook, *Born to Die, Disease and New World Conquest, 1492–1650* (Cambridge: Cambridge University Press, 1998); and Massimo Livi-Bacci, *Conquest: The Destruction of the American Indios*, trans. Carl Ipsen (Cambridge: Polity Press, 2008).

45 José de Acosta, *Natural and Moral History of the Indies*, trans. Frances López-Morillas (Durham: Duke University Press, 2002 [1590]), 143–144 (my emphasis). Colonial writers continued to draw on these ideas to explain mortality among Amerindians well into the eighteenth century; see for instance *Mercurio Peruano de Historia, Literatura, y Noticias Públicas*, vol. 5 (Lima, n.p., 1792), 195–196.

46 Cárdenas, *Problemas y secretos maravillosos*, 202–203 (quotes). On sadness causing epidemic disease among Amerindians see also Martínez, *Reportorio de los tiempos*, 261.

47 Rodrigo de Vivero, 'Tratado ecónomico político, c. 1609', in *Documentos inéditos para la historia de España*, ed. M. Ballesteros Gaibrois, vol. 5: *Papeles de Indias* (Madrid: Editorial Maestre, 1947), 34. See also Juan Botero Benes, *Relaciones universales*, trans. Diego de Aguiar (Valladolid, n.p., 1603), 151; Lizárraga, *Descripción breve de toda la tierra del Perú*, 16, 64; and Hernando de Santillan, 'Relación del origen, descendencia, política y gobierno de los incas, c. 1563', in *Crónicas peruanas de interés indígena*, ed. Francisco Esteve Barba (Madrid: Biblioteca de Autores Españoles, 1968), 144.

48 Thomas Palmer, *An Essay on the Means How to Make our Travailes, into Forraine Countries, the More Profitable and Honourable* (London, n.p., 1606), 46–47; Vaughan, *Approved Directions for Health*, 8 (quote); Neale, *A Treatise of Direction*, 31–32; and Kupperman, 'Fear of Hot Climates in the Anglo-American Colonial Experience'.

49 Description of a voyage from Lisbon to the Island of São Thomé, c. 1540, in *Europeans in West Africa, 1450–1560*, ed. John William Blake, 2 vols. (London: Hakluyt Society, 1942), I:157.

50 Andrew Borde, *The Fyrst Boke of the Introduction of Knowledge*, ed. James Hogg, 2 vols. (Salzburg: Universität Salzburg, 1979 [1547]), II:88 (quote); James Howell, *Instructions for Forreine Travell* (London, n.p., 1642), IV, 5; and J.N. Hillgarth, *The Mirror of Spain, 1500–1700: The Formation of a Myth* (Ann Arbor: University of Michigan Press, 2002), 13, 16.

51 Joseph Hall, *Quo Vadis? A Censure of Travell* (London, n.p., 1617), 18 (my emphasis).

52 Alonso de Alcocer to Juan de Colonia, Mexico City, 10 Dec. 1577, in *Cartas privadas de emigrantes a Indias*, ed. Otte, 99. A *chapetonada* is a colloquial term for an illness specifically afflicting newly arrived Spaniards, or *chapetones*. Or see María Díaz to Inés Díaz, Mexico City, 31 Mar. 1577; and Diego Sedeño to Diego Gómez, Mexico City, 22 Nov. 1592, both in *Cartas privadas de emigrantes a Indias*, ed. Otte, 97, 121.

53 Baltasar Álamos de Barrientos, *Discurso político al rey Felipe III al comienzo de su reinado*, ed. Modesto Santos (Madrid: Anthropos, 1990 [1598]), 16. See also Juan de la Puente, *Tomo primero de la conveniencia de las dos monarquías católicas, la de la Iglesia Romana y la del Imperio Español, y defensa de la precedencia de los reyes católicos de España a todos los reyes del mundo* (Madrid, n.p., 1612), 363, 21; Francisco Hernández, *Antigüedades de la Nueva España (c. 1574)*, ed. Ascensión H. de León-Portilla (Madrid: Historia 16, 1986), 97; and María Elena Martínez, *Genealogical Fictions: Limpieza de Sangre, Religion, and Gender in Colonial Mexico* (Stanford: Stanford University Press, 2008), 201–202.

2 Jean Bodin and the idea of anachorism

Richard Spavin

Can the physical world determine such complex concepts as freedom and liberty? Surely not. But it was nonetheless through the language of climate that thinkers first asked about the differences in social and political phenomena. Why are men different from one area to the next? How does geography influence governmentality? Since Hippocrates and Aristotle (460–322 BC), 'climate theory' represented a medium for a comparative politics: societies are different because they evolve in response to distinct climatic features.

In the case of the sixteenth-century French jurist Jean Bodin (1530–1596), one of the first 'classics' of the discourse,[1] climatic determinism is ever-present. As many have pointed out, it appears in most of his important works, reintroducing an ancient mode of thought to a wide range of topics (history, politics, sorcery, natural theology) in early modern French philosophy. Throughout the age of discovery, the effects of climate were common currency for Europeans coming to terms with new civilisations and races. As David Livingstone has shown, the theory extends far into our geographical modernity.[2]

In their various guises, from the sixteenth to the twentieth centuries, geographical 'determinisms' are often regarded in either of two ways: they represent, on the one hand, a controversial ideological tool for organising human diversity or, on the other, a significant ecological precursor for thinking about the very real correlations between the physical and cultural worlds.[3] New research on the concept of the 'Anthropocene' and of our own anthropogenic geological age can often resurrect the physical and moral 'relationality' in the determinisms of the past.[4] Indeed, certain early modern philosophers like Bodin and Montesquieu or naturalists like Buffon are part of the *history* of space and place, early references in the field of 'human geography'. When speaking about climate's discursive origins, however, we can often get confused: are we reading representations of climates for themselves or for our own conceptual genealogies?

Over the course of his career, Jean Bodin will offer three distinct formulations of his so-called climate theory: first, in the *Methodus ad facilem historiarum cognitionem* (1566), then in his famous *Six livres de la République* (1576) and, finally, in his more obscure *Universae naturae theatrum* (1596). In this final treatise on natural theology, the great climatic system of the *Methodus*, where geographical zones coalesce with planets, humours and politics, breaks down: Bodin abandons the moral connection between humans and their environment. Places are thought of as distinct, and the environmental influences on living organisms are reduced to mere physical traits. While we might be tempted to read a heightened sense of scientific exactitude in this last approach to climate, such a neat interpretation of Bodin's trajectory is ultimately not the most accurate.

Rather than scientific progress and theoretical refinement, I will show how Bodin's revisions reveal an unwavering belief in the power of metaphor. Faced with the considerable social upheaval in the second half of sixteenth-century France, the jurist's philosophy uses the science of climate to express moral imperatives of politics which change over the course of a thirty-year period. By focussing on the rhetorical dimension of Bodin's climate theory, this chapter will help reevaluate an 'epistemological' bias in the history of geography: namely, Hippocrates' superiority over Aristotle, or science's debunking of rhetoric and ideology.[5] The idea of geographical 'ana*chor*ism', a play on its historical counterpart, the better-known 'ana*chron*ism', will demonstrate how a writer can use climate rhetorically, as a creative and self-reflexive idiom, aware of its erroneous yet strategic representation of space. After introducing some key structural features of climate theory, I will discuss how the philosopher's recourse to it evolves alongside the turbulent Wars of Religion (1566–1598) in the *Methodus,* the *Six livres de la République* and the *Theatrum*. In this final text, the absence of a systematic environmental determinism speaks to the true complexity of climate theory: the all-encompassing idea of God leads Bodin towards a seemingly more scientific but deeply metaphorical approach to geographical knowledge.

Parameters of 'climate theory'

When confronted with climatic determinisms and the explanations offered for psychological, social and political diversity throughout history, our first approach may be to consider either their scientific primitivism[6] or the biased and often xenophobic comparisons that are drawn between different cultures. Attributing laziness to African character, for example, because of the effect of heat on human constitution, not only evokes a defunct scientific theory but sounds like racism: a way of perverting science in order to justify moral interpretations of otherness. Indeed, whether they are advancing the ideologies of Victorian colonialism, German imperialism or the Spanish *conquista,* geographically inspired theories of race have a way of making cultural hierarchies more believable, more ingrained in the order of nature. They have the effect of making purely imaginary and fictitious ideas – like the very notion of a hierarchy – seem like unassailable truths. 'Climate theories' are very much 'rhetorical devices of persuasion'.[7]

This is why the geographical and anthropological value of Aristotle's climatic zones have long been criticised. His notion of Greece as an ideally situated region leaves no commentator under the illusion of objectivity. Aristotle's *milieus* seem too determined by their own *milieu* to be of any consequence. Born out of Alexander the Great's wars of conquest, his 'theory' corroborates a project of universal hegemony, a Greek supremacy. Both scientifically and morally, critics have called it a 'bad' usage of geography that translates a political ideology into theoretical language: the emperor's dream of unifying continents under Macedonian rule.[8]

Intellectual historians are often critical of such ethnocentrisms, preferring the descriptivism of Hippocrates. The moral teleology and political biases are less pronounced in the geographical knowledge we attribute to the physician. The treatise, *Airs, Waters and Places*, is more concerned with the causalities between environment and sickness than with those between latitude and morality. Moreover, the representation of geography in *Airs, Waters and Places* attempts to go beyond a binary (Asia/Europe) or tertiary (Asia/Greece/Europe) categorisation of weather and/or geography. Hippocrates considers a multiplicity of factors that the travelling physician must deal with:

> Whoever wishes to investigate medicine properly, should proceed thus: in the first place to consider the seasons of the year, and what effects each of them produces for they are not all alike, but differ much from themselves in regard to their changes. Then the winds, the hot and the cold, especially such as are common to all countries, and then such as are peculiar to each locality.[9]

The term 'each locality' is meaningful for a pre-scientific conception of human geography which is open to different contextual possibilities. The widening of geography's parameters frames the empirical developments to come: political society is largely out of reach of blunt geographical influences; social mores are more the result of customs (*nomoi*) than of physical factors (*phusis*).[10] Next to Hippocrates' nascent empiricism, Aristotle can sometimes be regarded as a foil of geography's true origins.

But it would be short-sighted and historically inaccurate to edit Hippocrates down to his most scientific characteristics. Like Aristotle, he reduces climate to a schematic and metonymical structure: Asia's temperate and stable climate is in opposition to Europe's seasonal variations. Hippocrates is also not immune from moral interpretations of climate and otherness. The very beginning of the treatise states that both the physical and moral character of men conform to the nature of their country. And like Aristotle, he expresses a preference for the climate he lives in. There would seem to be a secondary geographical determinism at play that holds true for many 'climate theories' to come: the physician from Kos, an island off the coast of Turkey, praises the Asian temperament, whose inhabitants are more inclined than Europeans to engage in philosophical thinking.

Is Hippocrates' treatise contradictory? If the first half of the work presents determinist laws that the second half is supposed to corroborate, the work does seem disjointed.[11] His normativism yields to a scientific descriptivism in which the variability of geographic detail trumps his climatic dualism.[12] But there is only contradiction if we persist in trying to uphold the old modernist illusion that 'rhetoric' and 'science' are truly opposed. If by 'rhetoric' we mean a particular ordering of knowledge into persuasive schemas, 'science' can never fully be out of its reach.

Regarding this subject, the late Fernand Hallyn's work on the scientific revolution has shown the presence of a 'deep rhetoric' within the most empirically inspired works of the seventeenth century. Here, the discursive strategies of the 'new science' were to distance themselves from rhetoric and dialectic, seen as arts of speech and not of thought.[13] For the likes of Galileo, Descartes and Bacon, the sciences are distinct from the incertitude of human affairs. Questions of morality and law are reduced, in the eyes of Galileo, to 'rhetorical flowers', which do not possess any fixed or rational status. But rhetoric's condemnation does not exclude its practice. Hallyn has shown convincingly, for example, the structural place of metonymy in Galilean causality or the hidden role of commonplaces in Bacon's natural history. We should not read scientific discourse too literally: a poetic reading leads to a better understanding of the scientific imagination of the time, of how hypotheses were created and upheld, which would remain largely inaccessible through epistemological analysis alone. Conversely, when speaking of the 'rhetoric' of climate theory, our analysis must extend beyond the narrow usage found in Aristotle in order to better understand the variety of ways science can lend itself to philosophy and its discursive strategies in the early modern period.

Science, geography and the political imaginary

If the proponents of early modern science were wary of the inexactitudes of rhetoric, political philosophy's modes of expression were also influenced by the shift away from metaphysics and moral interpretations of human nature. 'Climate theory', for instance, can be seen as a step towards a progressive political 'empiricism,' defining power in terms of social realities rather than religious, moral ends. In its modern, relativistic approach to legislation, authors who wrote about climate introduced a philosophy of power to the ground, in all of its variations and imperfections. In the fifth book of the *République*, Bodin begins the discussion of climate by asking what is 'particular to certain Commonwealths due to the diversity of peoples, in order to accommodate the form of public life to the nature of place, and human ordinances to the laws of nature'.[14] The discourse injects a dose of realism into political philosophy and has made many Renaissance scholars consider Bodin's geography to be a transition towards modernity, one which is more empirical, categorical and open to social diversity.[15] Governance cannot be everywhere the same, and no matter how absolute the sovereign may be, laws must be in relation to the physical conditions which influence, in the first place, the social order.

Still, this empirically inspired approach to politics does not in itself constitute a 'science'. Climate's role in political philosophy is still profoundly moral and figurative. Politics is never neutral. Relativism is less, in other words, a scientific appreciation of the diversity of peoples than it is an argument for the recognition of customs, the need for constitutional breaks of power or even, in Montaigne's case, an argument against the false authority that colonisers assume in the New World.[16] Climates are more illustrative and persuasive than constative and scientific. In this way, we can compare the argumentation of climate theory with the anatomical and medical imagery of John of Salisbury or Claude de Seyssel, who use the notion of the 'body' to help the monarch better understand his responsibility towards his nation. If, for the mediaeval philosopher John of Salisbury (c. 1120–1180), the king is like the 'head' of a human body, the metaphor illustrates not his distinction *from* but his incorporation *in* society and law.[17] Similarly, Claude de Seyssel, in the sixteenth century, also uses the body as a powerful image for representing the very abstract intricacies of politics and balances of power. The notion of the king's 'mystical body' is inspired by the medical theories of the time: the humours illustrate the state as a system of competing forces, a precarious balance that will inevitably evolve into sickness in the absence of certain political 'remedies'.[18]

The choice of metaphor is also less dependent on science than it is on context. Like the body, the weather also became a popular cliché in political rhetoric, especially in the seventeenth century, when Machiavellianism was adopted by the likes of Descartes, Richelieu and his publicists. For these absolutist thinkers, the king is necessarily above the law and not fixed by any moral code. In fact, he can change the moral code to which he adheres at will. Such reversals of position are likened to the mutability of the weather: its whims, tantrums and unpredictability. Subjects must endure the political regime like they helplessly endure the seasons. 'The constancy of Kings, and their loyalty', says Jérémie Ferrier (1576–1626), 'is not to never change affairs and feelings [. . .] no more than God's will shakes with the movements of the air and the changes of season'.[19] Here, the brief recourse to meteorology helps justify the employment of the *arcana imperii*, or secrets of empire, as a persuasive device, to promote the reason of state. Even Richelieu will use the diversity of times, places and persons to assert the incomparability of historical and geographical phenomena, placing the king and the state in moral obscurity.[20]

Within the political idiom of geography, there are many opposing arguments and dissensions. If societies are different, over time and place, power and governance can never be the same from one place to the next. In the climate theories of Bodin and Montesquieu, this remains true. But the same ideas of mutability and inconstancy are no longer limited to the *arcana imperii*. They are stretched out and identified with different and highly schematic geographical zones. The climates of the north have different political regimes than the climates of the south, but their respective governments are destined neither to unpredictability nor to absolute incomparability. The two are commensurable and come together to express universal truths in a geographical language that is distinctly more neutral and less controversial than religion, direct political contestation or excusals of tyranny. Climate theory goes to great lengths to develop its 'scientific' dimension, extending the metaphor to hide the moral argument, presenting it as part of the natural order of things.

Indeed, for Jean Bodin, the ability to think about politics outside the language of religion is perhaps one of the greatest attractions to climate theory. Geographical determinism helps express the limits of his theory of power without having to name religious identities or choose sides. As we know, Bodin's career is almost perfectly concomitant with the beginning and end of the French Wars of Religion; the turbulence of the period furnished his writing with undeniable motives for restoring order and religious unity.[21] But we have yet to fully understand how the civil war relates to his geographical language, which itself undergoes significant change over a period of mysterious religious posturing, ranging between pro-catholic, protestant sympathiser, religious pluralist and judaiser.[22] If the *République* constitutes a political climax in Bodin's career, I will show how the discourse of climate is a rhetorical tool that helps articulate the evolutions in his thought, either historical, juridical or theological.

Climate and history in the *Methodus*

The *Methodus* is Bodin's most ambitious attempt at devising a so-called 'climate theory'. The work is itself a peculiar amalgam of rhetoric and science that purports to unify all knowledge under the scope of 'history'.[23] The preface describes this historical 'method' as an all-encompassing endeavour, capable of revealing the finality of all things, in the most accessible and pleasing of ways. For Bodin, history erases the divisions between different forms of knowledge and relates the observation of the past with the prevision of the future.[24] The work belongs to a late strain of *artes historicae*, dating back to Polybius: the past possesses an exemplary force that Cicero would transform into the famous motto '*historia magistra vitae est*'. History, master of life.[25]

This humanistic celebration of history presents a particular vision of the past which is not conceived of in isolation but in its symbolic relationship to the present and the future. It is what Zachary Schiffman calls 'a living past' in order to explain how the Renaissance interacts with Antiquity. Raphael's famous fresco, *The School of Athens*, serves as an ideal example of how this concept of the past is founded on a utopic atemporality, annihilating the limitations of historical time and geographical space, by putting different historical figures, dating from different periods, together in the same *stoa*. Schiffman, in following the footsteps of Thomas M. Greene, will go on to discuss the creativity of such anachronisms. In the case of Petrarch, who writes letters to Cicero and Saint Augustine, the rhetorical blurring of the past and present has a reflexive function and a critical edge, in that the creative deployment of historical 'errors' leads to a better judgment of self.[26]

Schiffman situates Bodin at the end of the Renaissance, at a critical moment in history that begins to regard the past as distinct from the present, due in part to his rigorous 'classification' of space and time. For Schiffman, the geographical components of Bodin's history help constitute a new understanding of the past, as if the climatic zones drew borders between 'here' and 'there' and 'now' and 'then'. In other words, the critical reflex is to assign a certain objectivity to geography that history lacks: the highly organised *liber locorum rerum* is not comparable to the poetry and symbolism of the *artes historicae*. In this way, the recourse to a geographical history is, for Schiffman, a departure from the humanist paradigm towards a more 'literal' reading of the past.

Schiffman's analysis supposes that a 'creative' dialectic between past and present is not transferable to geographical discourse. I hope to show the opposite. If anachronistic creativity exists in historical art forms during the Renaissance, the same can be said about spatial analogy or geographical error: 'here' and 'there' are comparable, forming an 'ana*chor*ism', a rhetorical play on representations of space that a philosopher like Bodin uses to arrive at a better understanding of his own social and political context.

When we look closer into the structure of Bodin's climate in the *Methodus*, we can see creative anachorism at work: climatic extremes begin to enter into correlation with one another and challenge the arguments for pluralism and relativism. Rather than expressing a true sense of diversity, Bodin's climates are famously tertiary, held in an equilibrium that extends from the ground to the planets. Bodin describes this as follows:

> Saturn, of course, is said to be cold, Mars warm, Jupiter more moderate than either. The first presides over knowledge and those things which find their realization in solitary contemplation of the truth; the second, wisdom, which is embodied in action, embracing all virtues; the third, arts and fabrication, which depend upon skill and strength. The first pertains to the mind, the second to reason, the last to imagination.[27]

The correspondences between humans, terrestrial geography, and the cosmos are threefold. If we look at Saturn, described as 'cold,' this 'first' planet favours sciences and contemplation. If we turn to Mars, which is 'hot,' this is actually the 'third' planet, which favours industriousness, physical strength, virility. The 'second' planet represents temperateness; it is second because it is in-between and associated with prudence. However, we must also note the inversion of characteristics attributed to the planets and to the principal zones. When discussing the Scythians as an example from the climatic zones of the north, Bodin characterises them as industrious, full of vigour and military prowess, but weak in terms of intellect. They are however associated with a *hot* planet. The southern climate is regulated in a similar way: its inhabitants are the most pious, the most contemplative and the most intellectual but are weak physically. Their planet is the *cold* Saturn. Rather than constituting contradiction, these examples translate a Hippocratic equilibrium whereby weaknesses are counterbalanced by strengths; cold planets beget hot-blooded attributes and *vice versa*. According to this schema, the force of nature is actually an opposing reaction to any geographical extreme, a natural system of temperance or balance. This three-part schema, with two extremes and a temperate zone, has the consequence of idealising but also targeting the middle climate to which Bodin gives specific latitudinal coordinates. The temperate zone is where nature and geography are the same. It is no accident that France is situated within this ideal climate. With France, Bodin represents a geographical model for thinking about laws and their 'tempering' and harmonising motives.[28]

Bodin presents his climate theory in the following terms: 'May we find facts decreed by nature and not human institutions, stable facts that nothing may modify.' Like history, climate informs the present and the future. But Bodin's main objective in using climate this way is even more incisive, as it is charged with the role of constituting a 'natural history' which may 'judge' what Bodin calls 'institutional history'. This compels us to think more deeply about what the status of 'nature' truly is and what relationship it has to 'geography' and to legislation, the jurist's end-point. If the temperate zone is ideal, the periphery is regulated through positive laws that 'temper' its tendency towards the extreme. These laws depend on wise legislators to recognise the disparity and to reconnect geography to nature, as is exemplified by countries that excel in jurisprudence, such as France. Perfectibility is therefore displaced, away from the French: corrective laws do not need to be imposed on the French but on their neighbours, whose natural reliance on legislation is undeniable. This is why knowing about 'natural history' is a way of overcoming environmental obstacles and of changing men's 'nature' with 'great force or prolonged discipline' so that they do not 'dangerously regress to their primitive state'.[29] In showing how power must yield to society, he also shows how society must, in certain instances, yield to the law.

In order to fully understand how Bodin's description of the temperate zone relates back to the French context, we must distinguish it from the extremities. Indeed, underneath the apparent diversity of the globe, the temperate zone illustrates a universal balance: it reveals the institutional defects of the extremities and highlights a 'methodic' and 'disciplinary' need to correct natural imbalance. With respect to the centre, however, the perfectibility of the temperate zone, by its very idealism, is absent; it can only exert itself on institutional discrepancies, which stay very much implicit for the French context. If the temperate zone is used to correct the extremities of the poles, the northern and southern climates can only prefigure the solution for a French situation in which institutional history has departed from natural history: the very problem at the heart of the *Methodus* – an indictment of institutional and not natural history. The relationality that grounds the different expressions of nature interclimatically, between the three zones, reinforces an identification with a need for reform over two distinct yet related modes of perfectibility – one activated by natural imbalance, the other by historical error. My hypothesis, therefore, is that the implicit correctives of the temperate zone are found *inverted* in the images of north and south. In other words, are Bodin's climates hidden metaphors for the critique of French society?

France as extreme: the case of the *République*

One way of testing whether non-scientific criteria drive Bodin's theory of climate is to read what Bodin says about climate in other texts, especially those published around the climax of the civil wars, such as the *République*, published in 1576. The evolution of Bodin's thought, one decade after the publication of the *Methodus*, may shed light on how his geographical language articulates political theory. The period separating the two texts is punctuated by the Saint

Bartholomew's Day Massacre, which ended a period of truce and brought the conflict to the streets of Paris. While Bodin's 'absolutist turn' is not necessarily a reaction to this external event,[30] the political situation does seem to require more literal and practical solutions to the civil war than those expressed in the *Methodus*. Indeed, the self-reflexive nature of his climatic discourse will become more explicit in the *République*, giving further credence to the idea that his 'climate theory' is an adaptive language that reflects Bodin's political concerns more than it does perceived geographic 'realities'.

Climates, discussed in the fifth book of the *République*, serve to delineate the space in which a sovereign power may be exercised. They give power a suggested legislative form, describing the different situations on which positive laws should be based. This is the 'relativistic' dimension of the theory which stays intact between both works. Furthermore, we find the same passage from determinism to indeterminism in the *République* that we saw in the *Methodus*, according to which 'discipline' is still a key addition to climate theory in cases of natural inequality, where one must 'correct the nature' of certain peoples through the institution of laws.[31] But the climatic target of this correction changes in the *République*. France is no longer as ideal as she once was in the *Methodus*. The nature of the kingdom has shifted:

> And one mustn't doubt, that men who come from a mix of these two peoples, are not more perfect than one or the other. If we desire more cheer in the Spaniard, and more swiftness than he has; and more moderate actions and passions in the Frenchman like the Italian seems to have in both cases, because Italy is in the most temperate zone there is, between the Pole and the Equator [. . .][32]

The temperate zone still corresponds to the 'mixing' of the two extreme characters, a happy medium well suited for governance and political stability. But the character of France has changed. It is less 'moderate' with respect to the middle, now limited to the Italian territory. What may the reasons be for this change? If the geographic character of France is stable, history reveals a different 'temperament' that is no longer in concordance with its true 'climate'. We can see the same shift happening in the following passage:

> The people from the middle regions, who are more reasonable and less strong, resort to reason, judges and trials. This is why we can be certain that laws and formalities came from these middle folk, from Asia Minor (where the great orators and preachers were celebrated) from Greece, Italy and France, of which a certain poet said, *Gallia causidicos docuit facunda Britannos*: because it is not since today that France is full of trials: and whatever laws or ordinances we make to eradicate them, the nature of the people will always prevail: as it is always best to resolve disputes through trial, if possible, rather than through blades.[33]

Bodin explains the confusion happening within the French character in terms that oppose the natural and the present. The French character of 'now' perverts the original nature which settles disagreements by 'trials' and not with 'blades'. The 'Britanni', to whom the 'Gallia' used to teach about laws, now instruct, in an inverted symmetry, the disturbances of French society. Blades are, indeed, the northern weapon *par excellence*, the cornerstone of their society: 'And the Scythians, said Solinus, buried a beloved glaive in the earth, placing in force and in knives the origin of their actions, laws, religion and judgements'.[34] The nature of the north now provides the vocabulary for judging the situation in France, through metaphor and analogy. But if the *République* seems to confirm, ten years later, the metaphor in the *Methodus*, this is not to say that rhetoric reigns supreme in Bodin's geographical thought. Nor is 'climate' a purely rhetorical construct in premodern geography. The true nature of environmental determinism is its discursive ambiguity, the way it oscillates, for specific cultural reasons, between rhetoric and science.

Climates and natural theology: the *Theatrum*

Twenty years after the *République*, in 1596, Bodin published his *Universae naturae theatrum*, a work on the divine physics of nature. The object of this text, Bodin's last published work, is the 'natural body' in its purest form, as opposed to the 'artificial body' (of interest to mechanics and the arts), divided into a hierarchy of ten 'hypostases', from the simplest to the most complex (inorganic to celestial), in order to contemplate the wisdom of God's creation. Bodin structures the work in the form of a dialogue between a teacher, *Mystagogus*, and a disciple, *Theorus*. The treatise is a long series of questions and answers, whose order follows the same hierarchy of the ten different levels in nature. These are all clarified, to the best of Mystagogus's knowledge, as necessary consequences of God's free will. In this way, the *Theatrum* resembles another of Bodin's later texts, the *Paradoxon*, a dialogue between father and son about divine providence and its relationship to virtue. Both works were meant to address the spiritual and religious crisis following the end of the Wars of Religion: they were cast as 'weapons' against man's evil-doing and blasphemy, for which the civil conflict is understood to be his divine punishment.[35]

Throughout the *Theatrum*, natural science reinforces the philosopher's theological ambitions. If Bodin's own words tell us that the work is intended to ward off impiety – an underlying ambition throughout his career[36] – his geographical language underlines a radically different point of view with respect to the *Methodus* and the *République*, one that is more significant than a 'change in tone':[37] the breaking down of his climate theory is, in this text, a departure from juridical science and its rhetoric but also, and more importantly, a retraction of his previous geographical symbolism, something he will himself criticise as a sort of blasphemous 'error'.

As Ann Blair has shown, climate does make some punctual appearances in the *Theatrum*. It appears for example when discussing the effects of seasons on different animals' weight. Bodin explains how some animals fatten up in different seasons: deer gain more weight in the summer than they do in winter; cattle are bigger in autumn. Birds, dormice, bears and men put on weight in the winter.[38] Here, climate is divided into different factors that influence different species in unique ways: the abundance of food, mating rituals and temperature may all vary from season to season. Blair also cites a passage on agriculture, showing how Bodin uses climate to explain why some plants grow on their own in certain regions and why the same plants die when tended to in other regions.[39] The explanation Bodin provides is typically Hippocratic: in maintaining an equilibrium between man and his environment, God facilitated the access to food which would balance the internal and the external temperatures. This is why spices grow more abundantly in torrid regions, where man's blood runs colder. These foodstuffs are absent in the north where men do not need them, as per God's great wisdom. For Blair, climates do not have the same categorising virtues they could have where, as in the *Methodus*, they are ordered cosmologically, in tripartite fashion. In the later text, Bodin yields to a case-by-case understanding of diversity and shies away from a universal 'law of nature'. If climates become less paradigmatic in the *Theatrum*, Blair does not explain this as a sign of 'progress'. The type of philosophy that Bodin is practising seems to suffer from an overabundance of knowledge and commonplaces, collected in a random way: 'Overwhelmed by the diversity of factors in nature, Bodin does not look for general patterns, even though his climate theory of temperaments might have been applied to good effect in this instance'.[40] While this observation is true with regards to the state of knowledge at this particular stage in the Renaissance, it does not explain the evolution of climate in Bodin's thought or the complexities of his new approach to geography, which seems to mature through his piety.

'Climate theory', as it was practised in the *Methodus* and the *République*, is certainly absent from the *Theatrum*, but there are good reasons for its absence. A careful reading of how Bodin uses climate in this final text reveals not only an attention to particular situations but a reticence towards the moral connections his geography once nurtured. More specifically, climate is no longer associated with the diversity of customs it once explained and justified. If, in the *Methodus*, the overarching structure of the cosmos accounted for differences in social behaviours, this is no longer the case in the *Theatrum*. Bodin shows how men and their laws are very much left to their own devices:

> But how can one call a discipline that [field] which above all relies on the principle that 'error makes the law'? From this it is well enough understood why from the memory of all ages, times, and cities the ends of this discipline could not be established, since it depends on the judgment and errors of men, so that what some judge worthy of reward others will condemn to punishment. These [thoughts] certainly called me back from the collection I had started on laws, gathered from all the customs and institutions of almost all peoples and compared with one another by lengthy labor so that something certain could be established. I realized that all the edicts, decrees, laws of the peoples are left rashly to the judgment and passion of men unless they rely on the law of God, that is, the law of nature, to guide their blind steps as if in a labyrinth.[41]

This important passage, cited by Blair, could be compared to the climatic passages in the *Methodus* and the *République*, where the 'laws of nature' coalesce with the laws of climate, thereby creating a chain of cause and effect, linking human behaviour, through geography, to divine providence. In these texts, the basis of human diversity was not seen as 'error' but as a consequence of God's plan. In the *Theatrum*, however, the philosopher's efforts are devoted to establishing an opposition between the falsehood of human laws and the truth of God's will. This discredits the philosopher's attempts at persuading about political imperatives through the language of nature.

The change in the moral approach to geography is most evident in the *Paradoxon* (1596), where Bodin argues for the absence of a 'temperate' zone or a natural home for jurisprudence. The full, translated subtitle of the *Paradoxon* speaks for itself: 'That there is not one single virtue in mediocrity, nor in between two vices'. The idea of a happy medium, an ideal space between two poles, was the very crux of Bodin's climatic discourse from the 1560s to 1570s. At the end of his career, the notion is relegated to an Aristotelian error, whereby a virtue cannot be in opposition to more than one thing. Here, the case of a temperate zone that balances out two different, opposing extremes – the France he had idealised in the *Methodus* – is very much in question. When he discusses temperature in the *Paradoxon*, he confirms his new, uncompromising stance on virtue, namely that two things cannot be contrary to one:

> Hot is opposed to cold, & not to dry, white to black, & not to green, as it is with others: if it were otherwise the ruin and subversion of the world would follow: but if moral virtue was between two vices, two things would be contrary to one: because Aristotle says not only that virtue is contrary to vice, but that evil is also contrary to good, & consequently vices are contrary to virtues.[42]

It is impossible to find virtue between vices which are, as Aristotle correctly says, *infinite*, or to find the 'more or less' when considering the 'infinite variety in the diversity of places, times, peoples, & in all sorts of mutable circumstances'.[43] Commensurability between places is no longer tenable: 'climates' are not tertiary but boundless. What we would call 'contexts' are dependent on a combination of factors which increase diversity exponentially. Virtue is found through contemplating God and his absolute goodness, not within the vicissitudes of men.

In the *Theatrum*, therefore, Bodin's climatic 'theory' is no longer a means to express moral order. Climates are significantly truncated, limited not only to an explanation of purely physical attributes but discussed primarily in terms of animals. When asked about the effects of climate on animals, Mystagogus explains that coldness retracts the skin and increases warmth inside the body, contributing to larger size. Animals are larger in Europe than they are in Asia. When it comes to man, however, Bodin's logic falters. And it does so significantly, for very specific reasons: 'this is why men are more robust in summer than they are in winter'.[44] With this logical inconsistency, whereby heat has the effect of making men stronger, Bodin contradicts what he says about the same effect on animals. We can only conclude that the philosopher is trying to use climate to *distinguish* between species, when they could be brought together, as comparable entities. Indeed, Bodin is avoiding a discussion of universal truths which would lead to an understanding of natural law – what his jurisprudence had previously depended on. This is why the discussion is separated from other potential moral questions such as courage and intelligence, previously shared between climatic poles. In response to the question 'why is it that courageous animals have a small heart, & scared ones bigger hearts?', Mystagogus responds that it is because force, when restricted to small spaces, leaves the heart with greater violence.[45] Again, courage is limited to animals, dissociated from man, as well as from climatic influence. When discussing animals' robustness in northern climates, he leaves out the climatic explanation for the size of the heart, precisely because 'courage', within man, has a moral connotation, one that was used to explain man's propensity for violence and war in previous texts, as well as to denounce the social extremes of the Wars of Religion.

Like Bodin's discussion of animals, two other phenomena have the same function of moving comparisons away from a moral geography. Seasons, as we have already seen, concentrate the effects of temperature on the same, singular region, obscuring the potential for comparison between different zones. Gender serves a similar function. When discussing climate, Bodin's perspective often defaults to the differences between the sexes. The discussion of male and female libido has, for instance, recourse to a seasonal determinism:

> Th. Why is it that male animals are commonly more lustful in the morning & in the winter, & female ones more so in the summer and in the evening, more than in any another season? My. Because males languish when weakened by external heat, which makes them lose their internal heat; in a similar way, women are more debilitated by the cold in winter, especially because their nature is of a humid and cold complexion: but when the cold restrains the heat inside of men, & the heat of summer moderates the coldness of women, each sex is more disposed to the venerial act, according to the season, which is specific to their temperament.[46]

The *Theatrum*'s explanation of libido differs starkly from a similar discussion offered in the *Methodus*, where southern men are described as necessarily more lascivious and perverse than their northern counterparts in order to compensate for a lack of virility.[47] The comparison in the earlier text includes an obvious moral criticism, whereas the later text resorts to a universal distinction between the sexes. Even though the climatic division of the planet is still manifest, explaining why there are more women in the south than in the north (because men of weaker constitutions tend to have female children),[48] it is noteworthy than men are not differentiated morally over environmental distance or through character or temperament. Their differences are reduced to strengths and weaknesses on a physical level, and their moral differences are diluted into a generalised, demographical relationship to the other sex.

Bodin's palinode: 'climate theory' as blasphemy

If the *Theatrum* is a fairly typical compendium of humanistic commonplaces, it is also an apolitical profession of faith, which occupies a particular space in Bodin's oeuvre. Published in the year of the philosopher's death, the work reads as a final, philosophical testament. The idea of revision or retraction comes naturally to this sort of writing,[49] and although an author's final word cannot subsume all previous utterances, it is one that tries to. In this way, the text differs in a 'programmatic' and 'final' way from the *Methodus* and the *Republic*, meaning it is generically distinct and has specific goals:[50] the *Theatrum* seeks to persuade about divine providence and represents Bodin's departure from the world of jurisprudence. There is, therefore, a perceptible will to escape the political through theology. Climate, in particular, helps us grasp this motive for authorial revision.

The principal change in Bodin's climatic discourse can be found in the way he opposes divine will and geographical symbolism. When Bodin is at his most pious, the determinism of climate is reduced to banal observations on size so as to avoid the teleological trap of *indeterminism*: Bodin no longer speaks about the 'discipline' or 'training' required to overcome natural inequalities. There is no longer a 'temperate zone' that can make sense of perfectibility. Laws, as he says, are the results of error. The only corrective proposed in the *Theatrum* is the contemplation of God's greatness. If the motives behind this authorial revision are certainly complex, they are most discernible in the ways Bodin expresses causality and divine providence, which invalidate climatic representations of human law.

The natural theology of the *Theatrum*, structured through its different hypostases, prohibits shifts in the different levels of causality. Since God is represented as the supreme free will that creates the earth in all of its diversity, Bodin is careful to avoid positing a causal relationship between human physiology and human laws. Why is this? The *Theatrum* recognises a certain danger in the rhetorical expression of physical nature: any effect seen to result from geography (and God's creation) can be linked to divine providence. Consequently, the political and juridical theory of the *Methodus* and the *République* is no longer an *effect* of geography (or of God's will) but a *cause* of how geography is represented. In other words, geography becomes tautology. The moral imperatives of the moment inform how Bodin organises geographical knowledge. Herein lies the 'error' at the heart of Bodin's retraction. His final text denies the process of mirroring political criticism within the signs of the environment. When asked if 'the celestial bodies' are 'made for man', Mystagogus responds:

> That is like asking if men were made for beasts: there is an infallible rule that the end is nobler than those things which tend toward it. But who, except one who is insane, would think that terrestrial things were superior to heavenly ones, the elementary world to the stars, darkness to light, and effects to causes? The arrogant ineptitude of men, however, is involved in this error, that they think that the stars and heavenly bodies from which they derive incredible usefulness were made for them.[51]

The 'arrogance' that Bodin is criticising is a general fallacy in causation. Man cannot be made for beasts because animals can be domesticated by his will and knowledge. Animals cannot, then, hypothetically, misrepresent this causality in order to *justify* their own interests. This is the same perversion of the chain of being that we have seen in previous expressions of climate theory. In the *Methodus* and the *République*, climates are made to reflect political aspirations and criticisms: they are distorted so that they serve whatever political needs history puts in front of them. The 'error' of climate is the rhetorical reworking of the chain of being:

God (cause)
Geographical diversity (effect) (cause*)
Human diversity (effect*) (cause**)
Human laws (effect**)

In climatic determinism, it is often said that customs and laws are made out of human diversity (the effect of geographical diversity). But can geographical diversity be 'made for' laws? There is a subtle tautology in the *Methodus* and the *République* that leads to what has become for Bodin a theologically unacceptable consequence: when human diversity is the reflection or foil of the laws that we want to institute, those laws start reflecting our terrestrial concerns more than those of God, which are unknowable. This is the rhetorical step that is left out of the *Theatrum*, which underlines the theological error of a political appropriation of climate theory.

The *Theatrum* is therefore a paradoxical antidote to environmental rhetoric. Bodin reintroduces God, on the one hand, to counteract naturalistic explanations of nature that see it as a self-regulatory system, a necessary chain of physical causes, to which God no longer needs to attend. On the other hand, his *Theatrum* corrects distinct moralisations of climate whose interpretations of nature are a pretext for human interests. The semantic zeugma of 'temperance', seen as both temperature and law, is cut out of his final representation of the environment.

Does this mean that we should consider the *Theatrum* Bodin's most scientific text? Not really. Even though the philosopher seems to come close to a more nuanced understanding of climate, his climatic discourse must not be dissociated from its previous, metaphorical expressions. For Bodin, climate is more of a personalised signifier, a way of mapping the philosopher's authorial revisions throughout his career. These revisions were made in order to respond to ideological pressures (historical, political and theological) that changed over the course of the Wars of Religion. Taken together, these intratextual elements demonstrate the ambiguity of Bodin's climate theory: not so much a 'theory' but rather a highly unstable yet consistently metaphorical discursive tool.

Notes

1 See Lucian Boia, *The Weather in the Imagination*, trans. Roger Leverdier (London: Reaktion Books, 2005), 43; Marie-Dominique Couzinet and Jean-François Staszak, 'À quoi sert la "théorie des climats"? Éléments d'une histoire du déterminisme environnemental', *Corpus* 34 (1998): 13–14.

2 David N. Livingstone, 'The Moral Discourse of Climate: Historical Considerations on Race, Place and Virtue', *Journal of Historical Geography* 17, no. 4 (1991): 413–434.
3 For an overview of the discourse, see Clarence Glacken, *Traces on the Rhodian Shore: Nature and Culture in Western Thought from Ancient Times to the End of the Eighteenth Century* (Berkeley: University of California Press, 1967). On current trends in the discourse, see Adam Gopnik, 'Faces, Places, Spaces: The Renaissance of Geographical History', *The New Yorker*. http://www.newyorker.com/magazine/2012/10/29. Last accessed 10 February 2016.
4 Christophe Bonneuil and Jean-Baptiste Fressoz, *L'Évènement Anthropocène. La Terre, l'histoire et nous* (Paris: Seuil, 2013), 36. See also Dipesh Chakrabarty, 'The Climate of Our History: Four Theses', *Critical Inquiry* 35, no. 2 (2009): 197–222.
5 On Hippocrates and Aristotle and their respective inclusion and exclusion in the genesis of 'geography', see Jean-François Staszak, *La Géographie d'avant la géographie. Le climat chez Aristote et Hippocrate* (Paris: L'Harmattan, 1995).
6 Mario Pinna, 'Un Aperçu historique de "la théorie des climats"', *Annales de Géographie* 98, no. 547 (1998): 322–325.
7 Livingstone, 'The Moral Discourse of Climate', 566.
8 Frank Lestringant, 'Europe et théorie des climats dans la seconde moitié du XVIe siècle', in *Écrire le monde à la Renaissance: quinze études sur Rabelais, Postel, Bodin et la littérature géographique*, ed. Frank Lestringant (Caen: Paradigme, 1993), 258.
9 Hippocrates, 'On Airs, Waters, Places', in *The Genuine Works of Hippocrates*, ed. Francis Adams, 2 vols. (New York: William Wood and Company, 1886), 156.
10 Couzinet and Staszak, 'À quoi sert la "théorie des climats"?', 13–14.
11 Indeed, this may be partly due to the contested authorship of *Airs, Waters and Places*. For such debates, see Philip J. Van der Eijk, 'Airs Waters Places and On Sacred Disease: Two Different Religiosities?', *Hermes* 119, no. 2 (1991): 168–176; and Jacques Jouanna, *Hippocrate* (Paris: Fayard, 1992), esp. 85–105.
12 Staszak, *La Géographie d'avant la géographie*, 143.
13 Fernand Hallyn, 'Dialectique et rhétorique devant la "nouvelle science" du XVIe siècle', in *Histoire de la rhétorique dans l'Europe moderne*, ed. Marc Fumaroli (Paris: Presses Universitaires de France, 1999), 603. See also Fernand Hallyn, *The Poetic Structure of the World: Copernicus and Kepler* (New York: Zone Books, 1990).
14 Jean Bodin, *Les Six livres de la République*, ed. Christiane Frémont, Marie-Dominique Couzinet and Henri Rochais, 6 vols., vol. 5 (Paris: Fayard, 1986), 7. All translations from Bodin's *République* are my own.
15 See Julian H. Franklin, *Jean Bodin and the Sixteenth-Century Revolution in the Methodology of Law and History* (New York: Columbia University Press, 1963), 69; Paul Avis, *Foundations of Modern Historical Thought: Machiavelli to Vico* (Dover, NH: Croom Helm, 1986), 52.
16 Michel de Montaigne, 'Apologie de Raimond Sebond', in *Essais*, ed. Jean Balsamo, Michel Magnien and Catherine Magnien-Simonin (Paris: Gallimard, 2007), 610–611.
17 See Michel Senellart, *Machiavélisme et raison d'état XIIIe-XVIIIe siècle: suivi d'un choix de textes* (Paris: Presses Universitaires de France, 1989), 24.
18 Nannerl Keohane, *Philosophy and the State in France: The Renaissance to the Enlightenment* (Princeton, NJ: Princeton University Press, 1980), 16–17.
19 Jérémie Ferrier, *Response au Manifeste du sieur de Soubize*, cited by Arlette Jouanna, *Le Prince absolu. Apogée et déclin de l'imaginaire monarchique* (Paris: Gallimard, 2014), 81. My own translation.
20 'There is nothing more dangerous for the State than those who wish to govern kingdoms by maxims taken from books; they will ruin them this way, because the past has no bearing on the present and the constitution of times, places and persons is different' (Armand Jean du Plessis, Cardinal-Duke of Richelieu, *Testament politique de Richelieu*, ed. Françoise Hildesheimer [Paris: Champion, 1995], 213).
21 Julian H. Franklin, *Jean Bodin and the Rise of Absolutist Theory* (Cambridge: Cambridge University Press, 1973).

22 See Paul L. Rose, *Bodin and the Great God of Nature: The Moral and Religious Universe of a Judaiser* (Geneva: Droz, 1980). And also Philippe-Louis Joly, 'Jean Bodin', in *Remarques critiques sur le dictionnaire de Bayle*, 2 vols., vol. 1 (Paris: E. Ganeau, 1752), 220; and Jacqueline Boucher, 'L'incarcération de Jean Bodin pendant la troisième guerre de religion', *Nouvelle revue du seizième siècle* 1 (1983), 33–44.

23 Inspired by Pierre de la Ramée, the term 'methodus' came to designate an almost rhetorical usage. If an author presented an art or a science in a brief and clear format, according to a set of rules, it was said that he 'reduced the subject to a method': see Marie-Dominique Couzinet, *Histoire et méthode à la Renaissance: une lecture de la Methodus de Jean Bodin* (Paris: Vrin, 1996), 37.

24 Jean Bodin, *Method for the Easy Comprehension of History*, trans. Beatrice Reynolds (New York: Columbia University Press, 1969), 9.

25 See Anthony Grafton, *What Was History? The Art of History in Early Modern Europe* (Cambridge: Cambridge University Press, 2007).

26 Zachary Schiffman, *The Birth of the Past* (Baltimore, MD: Johns Hopkins University Press, 2012), 144.

27 Bodin, *Method for the Easy Comprehension of History*, 111.

28 Ibid., 112.

29 We are using Pierre Mesnard's wording here, which grasps more accurately the inequalities in nature. In Beatrice Reynold's translation, the 'going back to nature' is seen, in this sentence, almost as a positive, whereby characteristics go back to their 'pristine character'. But this is precisely the 'danger' for Bodin, because climate is not a unified, homogeneous entity. Going back to nature can be a negative outcome for certain climates, hence the use of peripheral and extreme zones. See Jean Bodin, *La Méthode de l'histoire*, trans. Pierre Mesnard (Paris: Les Belles Lettres, 1941), 68.

30 For a new interpretation of the 'absolutist turn', see Sara Miglietti's article, which finds evidence of this change in position before Saint Bartholomew's Day, 'Meaning in a Changing Context: Towards an Interdisciplinary Approach to Authorial Revision', *History of European Ideas* 40, no. 4 (2014): 474–494.

31 Bodin, *La République*, vol. 5, 12.

32 Ibid., 26.

33 Ibid., 37–38.

34 Ibid., 37.

35 See Paul L. Rose, 'Introduction to the *Paradoxon*', in Jean Bodin, *Selected Writings on Philosophy, Religion, and Politics*, ed. Paul L. Rose (Geneva: Droz, 1980), vii.

36 See Paul L. Rose, 'Bodin's Universe and Its Paradoxes: Some Problems in the Intellectual Biography of Jean Bodin', in *Politics and Society in Reformation Europe: Essays for Sir Geoffrey Elton on His Sixty-Fifth Birthday*, ed. E.I. Kouri and Tom Scott (New York: Macmillan, 1987), 266–288.

37 Ann Blair, *The Theater of Nature: Jean Bodin and Renaissance Science* (Princeton, NJ: Princeton University Press, 1997), 19.

38 Jean Bodin, *Le Théâtre de la nature*, trans. François de Fougerolles (Lyon: Jean Pillehotte, 1597), 546–547.

39 Ibid., 388–390.

40 Blair, *The Theater of Nature*, 106.

41 Bodin, 'Dedicatory Epistle', in *Universae naturae theatrum* (Frankfurt: Wechel, 1597), Sigs. 2v–3r, cited by A. Blair, *The Theater of Nature*, 20. Translation is A. Blair's.

42 Bodin, *Paradoxe*, in *Selected Writings on Philosophy, Religion, and Politics* (Geneva: Droz, 1980), 60. Translation is my own.

43 Ibid., 63.

44 Bodin, *Le Théâtre de la nature*, 547. All translations from Bodin's *Théâtre de la nature* are my own.

45 Ibid., 548.

46 Ibid., 556.

47 Bodin, *La Méthode de l'histoire*, 87.
48 Bodin, *Le Théâtre de la nature*, 558.
49 For professions of faith and matters of conscience in testamentary writing, see Philippe Ariès, *L'Homme devant la mort* (Paris: Seuil, 1977), 188–195.
50 On different types of authorial revision, see Miglietti, 'Meaning in a Changing Context', 477.
51 Bodin, *Le Théâtre de la nature*, 881.

3 Marshes as microclimates

Governing with the environment in early modern France

Raphaël Morera

Environmental regulation and the government of natural resources occupied an important place in the legislative programmes of early modern states. In general, early modern states sought to manage resources and direct how they could be exploited in order to avoid scarcity. They did this both for the public good and to promote specific interests.[1] A scarcity of wood, for example, would have been disastrous given its central importance in everything from furnishing the navy to providing heat.[2] In this regard, historians have produced stimulating accounts of the development of the governance of natural resources and have focussed on the role of the control of natural resources in the growth of states. They have shown that the control of natural resources was crucial in processes of state formation, as well as for the enrichment of elites and administrators.[3] In France, historians have convincingly demonstrated how the growth of the state shaped French environments and landscapes.[4] Thanks to these studies, the way the state governed the environment is now clearer than ever.

These narratives are concerned with people insofar as they are touched by some of the consequences of the territorial planning of the French state. For instance, people who lived next to forests have received most attention because they directly suffered from the changes involved with new management practices. In these schemes, environments seem to have been constructed next to or even against people. This is all the more striking given that it is now generally accepted that the French monarchy was deeply interested in controlling people and behaviour. For Jean Bodin, in the second third of the sixteenth century, the only wealth is men.[5] Thus, the quintessential French theorist of sovereignty stated that the monarchy's main purpose was to expand the nation's population and wealth. As Michel Foucault has demonstrated, these guiding principles did not change throughout the early modern period.[6] Until 1750, Foucault observes a global attempt to control population by strengthening administrative oversight. It was in this context that states introduced identity papers in order to control the mobility and activity of their subjects more and more efficiently.[7] In the second half of the eighteenth century, new forms of governance progressively emerged.[8] Liberal thought was based on an understanding of human behaviours, so that good government had to use general rules to influence behaviours and finally increase the prosperity of the country.

A new appreciation of the relationship between environmental management and the government of human behaviours in the early modern period may come from taking into account what is often called 'climate theory'.[9] Originating in Antiquity, climate theory is mostly characterised by the relationship established between a range of environmental conditions and human behaviour.[10] According to Hippocrates, the 'airs, waters, and places' in which somebody lived determined their mores and constitution. This theory was assimilated by Islamic geographers, who divided the world into seven climates, similarly to the Greeks.[11] After the discovery of the lost Greek tradition in the twelfth century and afterwards, climate theory was adopted and revived in Europe during the Renaissance, alongside the rediscovery of the Ancient philosophy and medicine.[12] The renaissance of climate theory was particularly influential in France, where Jean Bodin used it to describe and explain the division between European peoples.[13] As late as the eighteenth century, it formed a key part in Montesquieu's thought. Notions of environmental influence deeply influenced early modern political theory and contributed to the legitimation of political powers. In so far as climate theory stated that human mores and abilities depend on the environment in which people live, it helps us bridge the gap between the governance of nature and the governance of people.

Land reclamation and draining are perfect examples of this logic. Across the early modern centuries, political and economic powers tackled global demographic growth by enlarging the area of cultivated soils. Drainage and reclamation schemes were particularly prominent in the Netherlands and in northern Italy but also touched the whole European continent and North America.[14] The development of hydraulic techniques and massive investment enabled the reclamation of a huge quantity of new land. In France, just as in England, the role of the central state was crucial.[15] For the French monarchy, the Moulins' Edict (1566) was a turning point.[16] It drew a clear distinction between watercourses: those which were navigable or floatable belonged to the king, and all others remained in the hands of the landowners. In many ways, the Moulins' Edict was confirmed by the Waters and Forests ordinance of 1669 and could be considered the oldest law still applied in France.[17] From 1599 onwards, marshes had been subject to another edict for the draining of French swamps and lakes.[18] Initially, this edict gave Humphrey Bradley, an engineer from Brabant, the right to drain all the marshes of the kingdom. If necessary, he could expropriate land and get associated with anyone. Whilst the legal framework continued to evolve until the end of the eighteenth century, the 1599 edict laid the founding principles of drainage policy.

In this regard, climate theory calls into question our distinction between environmental policy and the more general goal of controlling people. Focusing on the French marshes, the hypothesis of this chapter is that the French monarchy not only managed the environment and controlled its subjects but also governed *with* the environment. By supporting the draining of the marshes, the monarchy aimed to alter French territory in order to favour settlement, to improve the soils and to boost agriculture. For the first time, the monarchy then launched a long-term policy to remove marshes from the kingdom. In this context, this chapter will argue that climate theory provided intellectual tools which legislators used to legitimise policy. First, to develop this idea, we will have to question the specific representation of wetlands as microclimates within broader climatic areas. From this, we will be able to understand how these conceptions were used by the monarchy in order to back its agrarian policy. Finally, we will estimate the effectiveness of these policies and how ideas were transformed into practice. Here climate theory can be seen as a common cultural reference which the monarchy used to lead effective policy.

Marginal territories: marshes, climate and microclimate

In the early modern period, climate theory had a double dimension.[19] On the one hand, climates corresponded to a division between different parts of the planet. From this perspective, climates were purely geographical, situated places, influencing men and activities all around the Earth. On the other hand, climates distinguished between human behaviours and mores and were a part of medical thought. Medicine used the idea of climates to explain diseases and natural properties of different human groups. To understand the place of marshes in the context of climate theories we must then consider both of these traditions. Indeed, contemporary thought understood marshes as both specific places and determinant influences on certain behaviours. Since Antiquity, marshes were thus depicted as unwholesome areas where people were weak and threatened by diseases.

For Hippocrates, human development is endangered by proximity of wetlands. In his treatise *Airs, Waters, Places*, written during the apogee of Classical Greece, Hippocrates constructed a *topos* regarding the inhabitants of marshlands which enjoyed long-lasting influence. He described people living along the River Phasis in negative terms: 'the waters which they drink are hot and stagnant, putrefied by the sun and swollen by the rains'.[20] This was partly explained by the flow of the Phasis, which was 'the most stagnant and most sluggish of all rivers'.[21] As a consequence, 'the fruits that grow in this country are all stunted, flabby and imperfect, owing to the excess of water, and for this reason they do not ripen'.[22] Subsequently, Hippocrates explained the constitution of the Phasians as a result of their environment, in particular the marshes they inhabited:

> For these causes, therefore, the physique of the Phasians is different from that of other folk. They are tall in stature, and of a gross habit of body, while neither joint nor vein is visible. Their complexion is yellowish, as though they suffered from jaundice. Of all men they have the deepest voice, because the air they breathe is not clear, but moist and turbid. They are by nature disinclined for physical fatigue.[23]

As early as the fifth century BC, a causal relationship between the natural environment and human health and mores was postulated. Documentation is far too lacunar to suppose as to the concrete implementation of these theories into specific action on the ground, but the Greeks did launch massive draining projects in the following centuries, and hydraulic engineering became a key part of Greek urban infrastructure projects.[24] A few centuries later, Latin authors reinforced this negative view of marshes. Vitruvius made a clear distinction between marshes which were drained for cultivation and others which endangered people, their masters and their masters' property: 'Those places, however, which have stagnant marshes, and lack flowing outlets, whether rivers or by dykes, like the Pontine marshes, by standing become foul and send forth heavy and pestilent moisture.'[25]

At the same time, Varro and Columella contributed to refining these theories. Rather than dealing with people living in wetlands as a whole, they considered specific groups living in the marshes as a part of a wider population. For Columella, it was obvious that any investor or landowner must avoid marshes for agricultural exploitation or the construction of a villa. Columella advised against building any farms or houses near a marsh:

> for the former throws off a baneful stench in hot weather and breeds insects armed with annoying stings, which attack us in dense swarms; then too it sends forth plagues of swimming and crawling things deprived of their winter moisture and infected with poison by the mud and decaying filth, from which are often contracted mysterious diseases whose causes are even beyond the understanding of physicians; and at every season of the year rust and dampness play havoc with farm implements and equipment, and with unstored and stored produce.[26]

For the Romans, marshes were repellent environments within a favourable climate. In that sense, marshes were considered as microclimates, or specific areas where mores and human behaviours were different from those of the surrounding regions. Roman authors introduced new elements: for them, it was possible to modify the microclimate of marshes thanks to human technology. For example, colonies of veterans could plan and equip them for human habitation.[27] Human intervention entered the field of reflection: humans could modify nature, thus shaping an artificial environment, and in the same movement act on human behaviours and health. But even if humanity was empowered to modify behaviour through the environment in this scenario, the link between mores and ways of life on the one hand and nature and environment on the other hand remained crucial in the reasoning.

During the Middle Ages, this negative view of marshes appears to have receded to some extent, as it no longer inspired concrete environmental practice.[28] In the Renaissance, however, classical representations were mobilised again by several authors. Indeed, by this time the scale of desiccation had so massively increased that it needed to be justified. In his work as an architect, Leon Battista Alberti, a man interested in neither agriculture nor husbandry, devoted teeming pages of scholarly study to drainage as he grappled with the problem of water and marshes.[29] Moreover, natural conditions, air and water quality, sunshine and wind directions all governed the choice of places where human activities should be concentrated and towns should be constructed. Alberti considered many scenarios, and his opinion about valley bottoms was unusually sharp:

> [T]here is no site less suitable or seemly for any building whatsoever than one that is hidden away in some valley; for (to pass over such obvious reasons as that, being out of sight, it can enjoy no honor, while being denied the delights of a view, it can have no charm) it will inevitably suffer the ruinous torrents of rain and swirling floods; by absorbing too much damp, it will always rot; and it will constantly exhale earthy mists so damaging to man's health. In such a place no man could retain any strength, as the spirit wilts, nor any body show stamina, as its joints are weakened; mold will grow on books; tools will rust away, and everything in the stores will decay from excess of moisture, until it is all ruined. Furthermore, should the sun break through, reflected rays would cause the heat to grow more intense, but if kept out, the shade will make the air coarse and stagnant. What is more, should the wind penetrate as far as that, it would only rage with more violence and fury by being forced through fixed channels, but should it not reach there, the air would become as thick as mud. It would not be unfair then to consider such a valley as a puddle or a stagnant pool of air.[30]

The place of marshes in climate theory became more complex during the sixteenth century as land drainage developed significantly in Europe. The influences of both Neoplatonism and the Old Testament were central to this development. Roughly, Neoplatonism stated that men had to reveal the divine order which actually ruled the world but was hidden in nature.[31] With specific regard to drainage works, Neoplatonism was more influential in the Catholic countries, particularly Italy, than it was in Protestant countries. For instance, in the Netherlands, the story of Genesis played an important justificatory role, affirming that 'the water under the sky be gathered to one place [so that] dry ground appear'.[32] In both traditions, humans were empowered to act on nature in order to modify it and to accomplish God's will by shaping the environment.

Each of these conceptions was conditioned by specific political contexts and aimed to strengthen political powers. In the context of the Wars of Religion and the deep divisions in French society, Bodin could not adopt such an argument so easily. His primary goal was to reinforce the king's authority in order to overcome religious fragmentation between Catholics and Protestants. To do this, Bodin forged a notion of sovereignty to justify the concentration of powers in the hands of the king. In order to pacify the kingdom, Bodin proposed to develop harmony and friendship in the political society.[33] Taking into account that the French monarchy was too weak to lead an environmental policy, Bodin stayed faithful to the global and inherited model of climate theory.[34] Bodin used it to explain the political and cultural diversity within Europe and differences between Europe and other continents. Bodin thought that societies were able to change their mores and associated customs by adopting new ethical discipline:

> Let us first speak of the nature of the people of the North and South, and then of the East, and West, and the difference between the mountainers & those that live in valleys, or in moorish places, or that are subject to violent winds: then will we shew how much discipline may change the nature and disposition of men, rejecting the opinions of *Polybius*, and *Galen*, who held, That the country and nature of the place did rule necessarily in the manners of men.[35]

Following these premises, Bodin elaborated his thinking on marshes. Bodin divided people into five great categories. He considered that inside each of them, some elements could modify the global behaviour and properties of a specific climate. Wind, mountains, the littoral and marshes thus explained exceptions within a particular climate. In this schema, Bodin compared the characters of peoples who lived in mountains to the characters of those who lived around marshes. For him, people who lived in these unfertile areas such as marshes were more 'careful, vigilant and industrious' than others.[36] In arguing this, Bodin qualified the inherited theory. Although he recognised the hostility of swamps, he asserted that their inhabitants, because they had to cope with and overcome such difficulties, were bound to develop specific abilities and to become cleverer. Inside each broader climate, marshes, like other geographical features, thus constituted microclimates.

One hundred and seventy years later, Montesquieu also used climate theory to explain the world.[37] The authors of the *Encyclopédie* estimated that he had exaggerated its positions and that he was far too harsh.[38] Actually, Montesquieu substantially developed and systematised climate theory. Following Bodin, he also depicted marshes as specific areas within larger climates. Like Bodin before him, Montesquieu intensively read classical authors to justify his ideas, but he developed new and interesting points. Montesquieu was more obsessed by the development of the state and its legal order than was Bodin. Considering the state, Montesquieu drew an opposition between areas with or without states, between agrarian and nomadic societies, and eventually between civilization and barbarians. Regarding Persia, Montesquieu thus wrote, 'Men, by their care and their good laws, have made the earth more fit to be their home. We see rivers flowing where there were lakes and marshes; it is a good that nature did not make, but which is maintained by nature.'[39]

The conquest of land against water is simply assimilated to the reinforcement of the central state. In this way, Montesquieu acknowledged that, just like the Persians, Tartarians did not have swamps in their country. He explained this contradictory fact by their climate, defined as weather, which he supposed to be too cold.[40] Moreover, for Montesquieu the existence of marshes threatened the state and could be seen as a sign of state failure. Thus, when nations did not cultivate the soils 'their country is ordinarily full of forests; and as men have not dug canals for water, it is filled with marshes where each band camps and forms a small nation'.[41] As a rule, marshes were linked to a kind of mismanagement of waters and consequently seen as rebel territories beyond the authority of a central power. In a way, marshes embodied resistance to the state. Beyond this opposition, it is possible to see a clear distinction between a settled way of life and nomadism. The permanence of swamps was conceived as a lack of cultivation and abandonment. It was also a sign of bad government: people were seen to be poorly guided and controlled, uncompelled to cultivate marshes through drainage. Likewise, nomadism was equated with political division and tribal in-fighting. Thus, from a different perspective than Bodin, Montesquieu also surmised that marshes were microclimates and very specific parts within wider areas.

Climate theory was thus very ambivalent. On first reading, it appears to have stated a permanent division between regions and peoples. Conceived during Antiquity, it carried immemorial bias and prejudices backed by means of undisputed authorities. Whereas political institutions lacked the tools to apprehend social reality, climate theory provided a global framework with which to categorise it. On the other hand, uses of climate theory sharply changed between the sixteenth and the eighteenth centuries. Whereas Bodin considered that humans could only act on their behaviours, Montesquieu stated that it was possible to change climate by human planning in order to promote a social pattern. However, in both traditions, the role of the state was determining: to rule population as well as to promote engineering. The relationship between climates and societies was conceived of as complex and dynamic, which is why climate theory encouraged political action and could be used as a legitimising discourse by the state power.

Changing climates: an agrarian policy

Climate theory began to be used as a political theory as early as the Middle Ages. Thus, Thomas Aquinas distinguished northern climates, where people are strong, from southern climates, where people are cleverest.[42] At the beginning of the fourteenth century, the French jurist Pierre Dubois justified French participation in the Crusades using climate theory.[43] Dubois stated that the French were the people most capable of recovering Jerusalem and taking care of the Holy City. For Dubois, France was the best qualified kingdom to lead the Crusades because its climate was the best. Moreover, the French king had to raise and educate elites needed in Palestine and considered the French climate to be favourable to nurturing talent. Another author, Jean de Jandun, used the climate theory in the context of the *translatio studii*.[44] The *translatio studii* theory stated that knowledge had circulated from Athens to Rome and then to Paris, where the Sorbonne was considered as the inheritor of Ancient science. Thus, Jandun, in his *Tractatus de laudibus Parisius*, written in 1323, claimed that the medieval French climate was equivalent to the classical Athenian climate.[45]

Climate theory was used to reinforce royal authority whilst the French monarchy developed a very powerful monarchically centred ideology. In this ideology the king was sacred and held his power from God. In an organicist framework, he was said to be the head of a society thought of as a body inside of which each constituent element had a specific task to fulfil. The king's job was to rule this body and to check that everyone respected his own role so that the social order could be protected. Even though peasants were bound by feudal bonds, they had a specific purpose in this context.[46] They had to sustain and to feed the wider society. Mathieu Arnoux has recently demonstrated that the skills and work of peasants were promoted because they were the source of medieval economic growth. Nevertheless, in most cases, they ploughed land they did not own. Their social world extended only as far as the community to which they belonged. To step outside of this framework was to be ostracised. In this sense, the legal and social orders underpinning the French monarchy were founded on settlement and the promotion of agriculture.[47] From the sixteenth to the eighteenth century, the French population remained overwhelmingly rural. On the eve of the French Revolution, peasants still made up 87 per cent of the population, and the shift to 50 per cent only happened in the twentieth century.[48] Furthermore, urban growth mostly depended on migration from the countryside: urban demographics were too weak to provide any surplus capable to encourage urbanization.

The French king was thus considered the patriarch of the nation. As a father, he had to provide his subjects with the ability to subsist, to support agriculture, and provide relief in corn in case of shortages. Pierre Charron, a philosopher and moralist writing in the second half of the sixteenth century, expressed this duty in political terms:

> The second means to acquire benevolence is beneficence, I mean to all first, including poor people, by providence and good police, whereby wheat and all things necessary to sustain this life are not lacking, but be in good proportion, nay abound if it is possible; that costliness does not worry subjects. Since poor people only care about the public for that issue, *vulgo una ex republicata annonae cura* [Tacitus].[49]

For Charron, securing the people's well-being was not a purpose in itself but a means to ensure their fidelity. Charron estimated that people were not able to comprehend and align with the imperatives of the public good. Providing for the people was one way of governing, one that clearly structured monarchical ideology. Despite his reputation as the 'King of Wars', a century after Charron, Louis XIV enshrined himself in this strategy of legitimation.[50] The *Galeries des Glaces* mainly represented the war success of the Sun King, in particular the great event of his conquests of the 1660s and 1670s. However, out of thirty pictures, two were concerned with trade and assisting the people: the struggle against shortage and the importation of Polish wheat in 1662, and the 'junction of the seas' in 1667.[51] When he decided to liberalise the grain trade in 1763/1764, the minister Bertin raised a wave of opposition in Parliament and amongst the people and revived rumours of a famine plot.[52] According to them, the king had to control the grain trade in order to ensure that people could eat: the market could not succeed in such an important task.[53]

Ensuring that the needs of the people be met was not just a political imperative. It was also implicated in an economic philosophy in which population was the main basis of power. Throughout the early modern period, political economy thus equated the wealth of nations with demographic prosperity. But how could people be counted? How was it possible to estimate demographic trends? Before the invention of demography in the middle of the seventeenth century it was almost impossible. It was difficult to generalise about population trends and even harder to apply them in political calculations. In order to circumvent this problem, demographic growth or crisis was mostly read from the landscape. In the context of relatively static technological advances, increasing the amount of arable land meant setting more hands to work, which could only mean demographic vitality. Therefore, uncultivated land was land that was not contributing to the wealth of the kingdom.

Aside from the second part of the reign of Louis XIV, the French monarchy continuously supported drainage investment in order to transform wetlands into arable and pasturable soils from the 1560s to the Revolution. The first initiatives can be traced back to the end of the sixteenth century and were consolidated in the edict of 1599. This text allowed Humphrey Bradley, an engineer native of Brabant, to drain all French lakes and marshes. Faced with the scale of the works and their dissemination, Bradley quickly formed a company in association with members of the aristocracy loyal to Henry IV and to his prime minister, Sully. After their death, Bradley's company sided with Richelieu and then Mazarin. Monarchical support for draining was essentially judicial: in case any opposition to drainage came to trial, the State Council served as the only competent court in which cases would be heard. This strategy of judicial support prevailed until the middle of the eighteenth century, after which it began to evolve. Under the edict of 1599, the challenge was not to strengthen the central power through the enrichment of its elites but to back local initiatives. In pursuance of these aims, the process was more and more systematised and ruled by the administration.

In this respect, the monarchy's strategy is similar to the promotion of ideas of climate defined by Bodin and Montesquieu. France was characterised by abundantly cultivated land, with a population able to work and be disciplined. Within this vision of France, marshes constituted a microclimate, highly differentiated from the main royal image of the French climate. They embodied and symbolised a failure of royal authority. Marshes embodied the antithesis of the monarchical ideal of France: an under-populated and under-cultivated land. To this extent, the promotion of drainage was underpinned by ideas of climate: it aimed to change the mores of particular territories. The edict of 1599 remained the legal basis for all land drainage acts until the French Revolution. Their principles were never reformulated but instead clarified or adapted to local situations. It stated:

> There is in Our Kingdom a great quantity of marshes and swamps flooded and undertaken by water, and almost useless, and of little profit, which hold several countries as desert and uninhabited, and incommode neighbouring countries, both because of their bad vapours and exhalations, and because they make travel difficult and dangerous; which marshes and swamps, once desiccated will serve parts in arable soil and parts in meadows.[54]

Strikingly, the edict of 1599 mobilised most of the *topoi* expounded in climate theory. Drainage was justified by economic, demographic and territorial arguments. Indeed, they were designed to help populate deserted areas, to increase the king's power and to develop the kingdom by the growth of cultivated soils. They finally unified, that is to say built up, the country by easing inland transport. Point by point, the edict tackled all the negative aspects of the marshes. The struggle against diseases and pestilence was not stressed, but it was a part of the dynamic of settlement promoted in the drainage projects. The conquest and improvement of marshes thus aimed to enforce a cultural as well as a socio-political order on natural sites. The ambitions of the monarchy were clearly related to desires to manipulate climate and can be compared to colonial processes. In 1639, rights granted to first investors had to be renewed and reformulated because the investors had changed. The principles of the first edict were confirmed and enshrined in a new declaration, which directly drew an analogy between the draining, that is to say the conquest of new lands, and the colonial process. The declaration stated:

> As the late King Henry the Great of glorious memory, our honoured Lord and Father, had well recognised that what the land produces within the extent under his rule, (so useful to his subjects and necessary to his neighbours) served him more than the India to the Princes who take advantage of it.[55]

The edict thus made an explicit link between extra-European and internal colonisation. French adventurers in America shared a common approach to the climate: they also wanted to change it. For the French settled in New France, the cold and wet climate of Québec was an obstacle to the success of colonisation. In that context, authors such as Marc Lescarbot and Pierre Biard propagated the idea that settlement, by developing agriculture, would rapidly modify the hostile climate of French Canada.[56] In the colonial process, climate was considered in both its cosmographic and geographic and moral and spiritual dimensions.

Just as in the French marshes, colonisation and the development of agriculture in Québec would contribute to the modification of climate and the transformation of inhospitable areas into habitable and profitable soils. This line of argument continued to be influential until the middle of the eighteenth century. In 1702, an edict regarding drainage in Languedoc (Southern France) continued this tradition.[57] Still in use in the 1740s, it stated that poverty in Languedoc

> comes from the scarcity of fodder, and that the places which could give more abundant fodder are occupied by waters, remain swamps in nature, and will only produce reeds inconvenient to cattle pasture. But, once desiccated they will turn into pasture and meadows which would furnish the inhabitants with food for many cattle, enable them to cultivate their land and even to use it in trade.

The edict aimed to change the inhabitants' way of life, imposing new cultures of cultivation and husbandry which would directly impact on the environment. With regards to Languedoc, the text stressed the growth of pasturable lands in order to promote dairy products suitable for trade, despite Southern France being ill suited to dairy farming. Less abruptly than texts of the seventeenth century, the edict of 1702 reproduced their justifications. It aimed to extend cultural practices from the north of France corresponding to vast plains and a temperate climate to Mediterranean France, where the natural conditions were completely different. In this regard, promoting draining was underpinned by the desire to change and eradicate local environmental specificity.

Montesquieu's work constitutes the apogee of climate theory during the early modern times. Even though climate theory was intensively mobilised during the nineteenth century, Montesquieu was the last great author of the eighteenth century to use it in order to explain the differences between peoples. As early as the end of the seventeenth century, the term 'climate' began to be more dedicated to geographic and cosmographic issues than to mores and human behaviours. In that sense, in the second half of the eighteenth century, the rise of political economy and physiocracy provided new foundations for drainage and land reclamation policy. According to the physiocrats, agricultural labour remained the foundation of wealth, and thus the growth of corn production was the absolute priority.[58] To that extent, physiocracy and agronomy were strongly linked in order to improve the yields, and agricultural innovation and land reclamation were thus encouraged. Nevertheless, in the physiocratic model, the growth of production was bound to the liberalisation of the grain trade to better distribute grain and avoid famines. In that sense, drainage was no longer seen as a way to propagate 'civilisation' but was more and more conceived as a way to promote the individual interest of landowners and, through them, to enrich society.

Generally speaking, the physiocratic theories deeply modified the justifications of the legal framework of drainage. The goal of drainage was still the promotion of economic growth and the enrichment of the country, but this was now to be done through the promotion of the individual interest of landowners. From this point of view, the royal declaration of 1764 is very significant. Over eight pages, it restated the legal framework elaborated since the end of the sixteenth century and paid tribute to the glorious kings who initiated drainage. But it was no longer a question of whether to give a monopoly to a specific desiccator or to a company, as the 1599 edict had done. The ordinance states that the legal power to drain land should 'be free and allowed [. . .] to all seigniors and landowners of marshes and flooded lands, commonly with those who would have taken or would take emphyteutic[59] leases or in perpetuity, to drain these marshes and flooded lands'.[60] The goal is no longer to modify the climate or to populate a region but to encourage individual enterprise and permit the local promotion of drainage. This declaration applied to a broad spectrum of people, since neither the lordship nor the ownership of land are required to launch works – having use of the land and a simple lease was sufficient. As a result, these dispensations considerably eased the administration of the drainage schemes.[61] After 1764, a promoter only had to declare to the intendant his intention to drain his marshes. Thus, the king's representative just had to control and to grant a permit. The liberal turn considerably changed the way the state justified drainage: it became purely a matter of business.

This important shift of the ideology could be understood as an abandonment of climate theory. But such a conclusion would be far too extreme: under the varnish of liberal thought the negative view of marshes survived. The physiocrats had in fact assimilated the negative view of wetlands elaborated during antiquity. In 1791, under the rule of the constitutional monarchy, a law was adopted to encourage draining all over the French territories. As in 1764, the goal was to free local investors and to support all the initiatives. But in this law, marshes were considered in the same way as in the sixteenth century. The law specifically targeted the growth of the population through the struggle against noxious marshes. A further law of 1807 drew a strong link between under-population and the persistence of marshes. In the context of the First Empire, the law emphasised the role of the state, which directly managed the huge works of Arles or Rochefort. Despite the rhetoric of liberalism, climate theory continued to influence and even to structure how wetlands were conceived.[62]

A concrete ideology?

Theoretical and legal texts give access to the stated intentions of political powers and to the intellectual basis of their decisions and actions. But they do not provide any information about the consequences of these intentions and ideas and their realisation in schemes on the ground. Beyond the statements of the monarchy and its theorists, important questions remain: were environmental projects an effective way to conduct government? Did the use of climate theory and its application in projects really change the lives of the people? To answer these questions, it is necessary to analyse what was undertaken on the ground. From 1599 to the French Revolution and beyond, laws paved the way for a huge wave of drainage. Unfortunately, it is almost impossible to give an accurate estimation of the extent of the surface effectively drained and newly devoted to agriculture as a result of new water management schemes. In the seventeenth century, the right to drain was in the hands of very small aristocratic elites, so that it is possible to assert that at least 15,000 hectares of marshes were drained between 1599 and the 1660s. Each of these works had a very deep local impact, modifying local environments and settling new populations.

The marshes of Petit-Poitou had first been drained by Cistercian monks during the Middle Ages. However, these drainage works were abandoned during the sixteenth century, partly because of the destabilisation of monastic life brought about by the Wars of Religion.[63] Supported by successive French Prime Ministers Richelieu and Mazarin, drainage works were resumed in the early 1640s. They consisted in renewing the medieval equipment and building new sluices to protect the area and ensure flow. These works required massive investments which were enabled by the commitment of Jean Hoeufft, the representative of the United Provinces in France, who exerted considerable influence in Paris. The account books, held by Octavio de Strada, the local agent of Jean Hoeufft, depict the progress of the works. Here the conquest of the marshes was realised by workers coming from Auvergne, where Strada owned a large domain, itself obtained through the drainage of a lake at Sarliève. These works were achieved by a transfer of population, which is unfortunately impossible to quantify. The presence of people from Auvergne's is clear from the place names of the drained area: several farms created by the drainage carry the name of villages surrounding Clermont. Investors built several farms all over the former marshes. They were conceived as exploitation farms and were based on the new arable soils. Still visible nowadays, they embodied the real goal of the draining: to settle a population in a former 'desert' in order to change waste lands into productive soils. Within only five years, desiccators deeply modified Petit-Poitou's marshes. Drained thanks to efficient hydraulic equipment, they launched a long-term programme of conquests of the French Atlantic littoral. Rich and productive meadows and fields whose production was oriented to European trade thus succeeded impenetrable swamps. Investors' archives enable us to piece together the agricultural models pursued in the drained areas. The ambition was not to encourage innovations but only to increase the quantity of cultivated soils. In this scheme, the traditional system of sheep breeding and grain farming largely subsisted, while cattle breeding remained limited. The development of equine breeding was the only innovation in the farming model.

The draining of the Arles marshes, in the Rhone delta, was also achieved in the 1640s and bore out the same logic. Again, Jean Hoeufft and Octavio de Strada were the main leaders of this project. After overcoming strong opposition thanks to the intervention of the state, they were able to cultivate the whole left side of the Rhone, from Arles to the Sea. Unlike in the Petit-Poitou marshes, no documents shed light on the new agricultural strategies initiated by the drainage. In Arles, the investors did not instigate a building programme, so that the hydraulic grid is the only vestige of the works visible in the landscape. At the local scale, the Arlesian drainage works nevertheless provoked a quick influx of population. In 1647, the Cathedral chapter of Saint-Trophime attempted to appeal to the church court against the desiccators. The case was finally judged by the Parlement of Provence. What was in dispute? To encourage the draining, the king had exempted the new lands of all kinds of taxes, including the tithes. The success of the programme had attracted peasants who used to cultivate lands submitted to the tithes. In other words, the draining of the Arlesian swamps deprived Saint-Trophime of these most important resources, so the bishop sought compensation. From this point of view, the Arlesian works were part of the same logic of those of Petit-Poitou. Environmental transformation through hydraulic engineering not only disrupted the agricultural practices but also modified a local climate, so that it became favourable to a settled way of life and was thus more amenable to state control.

In 1661, the beginning of the personal reign of Louis XIV marked a break in the policy of marshland reclamation. Colbert, the Sun King's most important minister, was less amenable to drainage, and the tax exemptions allowed to drainers were less acceptable while the country was penniless due to its continual involvement in European wars. Nevertheless, the dynamic of drainage recommenced in the eighteenth century, thanks to individual initiatives supported by royal disposition. This policy contributed to winning huge quantities of new land along the littoral as well as in inland valleys. The Charente and Poitou seashores and Maritime Flanders are the most famous examples, but the movement was widely spread throughout France. The main difference with the seventeenth century was the direct involvement of the local high aristocracy.

In 1599, Humphrey Bradley was granted the privilege to drain marshes and lakes and had to contract with Parisian investors to exercise it. Bradley was the king's man, and the company he established could act on the whole territory. In principle, the privileges of the eighteenth century were fundamentally different. Each of them was absolutely local and benefited the local aristocracy. Thus, in 1702 the edict for draining the Languedoc marshes was given to Marshal Duke of Noailles, cousin of Louis XIV, as a reward for his devotion to the king's services.[64] The method was exactly the same in Flanders. The so-called Moëres, a lake located around Dunkirk and first drained by the Spanish during the seventeenth century, was initially given to the marquise de Maisons and the marquis of Canillac. In consideration of the great difficulties in draining this place, the Moëres were given again to the Count of Hérouville in 1759.[65] The count of Hérouville was actually general lieutenant of the king's armies and general inspector of the king's infantry: receiving these lands made him lord of a large domain and was a means of rewarding him.

The drainage project of the Languedoc coastal plain faced the accumulated opposition of local communities, the nobility and the Church.[66] It failed, and drainage in this area was not achieved before the twentieth century. However, the several inquiries and disputes surrounding this project contributed to the broader dynamic of drainage. The judicial dispute led by the project's opponents was part of a controversy linked to climate theory, which totally reversed the usual arguments of the latter. First, the opponents asserted that drainage would increase fevers and would be pathogenic for them. They also feared that the drainage could cause the impoverishment and desertification of the country and that they might lose at least one year's revenue during the works. Conversely, the promoters of the projects argued that the works would not take more than one year and that they would be realised by workers coming from the mountains. In this model, the works would not engage the usual workforce of the local landowners but would instead contribute to populating the area.

Whereas during the seventeenth century the investors were capable of expropriating local landowners, this became more and more difficult in the eighteenth century. Studying the cases of Normandy and Languedoc, Jean-Laurent Rosenthal showed that feudal institutions caused the failure of great draining programmes.[67] According to him, the *Ordinance on Waters and Forests* of 1669 and the following ordinances of 1677 and 1699 imposed conditions which were impossible for any investors to respect. Under the rule of these laws, in order to drain a marsh, a lord had to negotiate with the community living around the area to share the lands. In any case, the lord or investor was supposed to hand over two thirds of the marshes to the community before draining it. One of the goals of Louis XIV's government was indeed to reinforce communities against seigniorial power in order to weaken this opposition force.[68] In that scheme, both the requirement to obtain the agreement of all the owners of a marsh and need to respect seigniorial power put up the cost of any programme. It was all the more difficult to obtain the agreement of the communities that used the marshes to provide a basic income to their poor.

Considering the cases studied by Rosenthal, these conclusions cannot be denied. In the case of the marshes of Languedoc, climate theory was thus used by both pro- and anti-drainage interests. Thus, clergymen from Nîmes and Beaucaire stated that 'the draining would cause a drought that would destroy everything within fifteen leagues: no one can ignore that, such vast marshes being the most important cause of rain, this [rain] would become very rare as soon as they were dried'.[69] To justify their fears, the drainage sceptics mentioned the pond of Pujaut, the drainage of which had ruined the surrounding areas and caused an important loss of lands. Further reappropriating climate theory, they also argued that drainage could contribute to the proliferation of mosquitos and pestilences. As an answer, the promoters of the project swept these arguments aside by reassessing the basic elements of climate theory: drainage contributed to the improvement of the air and did not cause any decrease in rainfall. Despite being able to support their claims with more than a century of experiments, they eventually failed.

Nevertheless, we should keep in mind that Rosenthal mainly used the archives of central administration and cases which had to be helped by the central government. In that sense, there is a kind of archival bias. Indeed, Rosenthal mainly focused on examples of failures which required the intervention of the state. As a matter of fact, records of local and successful draining projects are not kept in central archives. Furthermore, it is also remarkable that several attempts to drain marshes were successful during the eighteenth century. For instance, the swamps of Rochefort were almost totally drained in the second half of the eighteenth century.[70] The draining of Marais Poitevin was accelerated across the eighteenth century.[71] In certain cases, some enterprises were successful for only a short term before collapsing again because of technical limits more than institutional resistance. For example, the Intendant of Picardy, Chauvelin, strongly supported the cultivation of the area of Calais.[72] He succeeded in convincing the army and the local population to co-operate in this regard during the 1730s. But soon after the work was completed, water returned to the area because of a lack of manpower. More famously, the abandonment of the Arles marshes was explained by the neglect of foreign owners and an appeal for irrigation more than it was by feudal conservatism.[73] Even if much more research is still required to fully understand the success of land reclamation, several examples show that draining works could indeed be performed in the eighteenth century. In this context, the exact impact of the royal ordinance of 1764 encouraging all kinds of reclamation, by clearing and by drainage, has yet to be estimated.[74]

Conclusion

With regard to wetlands, marshes and lakes, early modern climate theory was not exactly the same as that of Classical Antiquity. Undoubtedly, political applications of climate theory had already been worked out by ancient and medieval authors such as Aristotle and Thomas Aquinas, and persisting ties with the Greco-Roman tradition contributed to the intellectual legitimation of the theory in Renaissance France. In the early modern period, a time of accelerated state formation, climate theory provided both a set of ideals to aspire to and the justification for those aspirations, and by firmly anchoring the practice of power in ancient thought, it glorified the monarchy in so doing. Moreover, the stereotypes that it conveyed were among the most powerful ways in which some of these issues were understood. They had an undeniable strength of conviction and were thus a lever for action. At a time when powers did not have any expert instruments at their disposal, climate theory could serve as powerful justification by furnishing state action with legitimate representations of wetlands. Finally, in the French case, climate theory was so influential because of its malleability and extraordinary adaptability. It could easily be adapted to the needs of powers and was thus a very useful ideology.[75]

Usually climate theory is criticised for its tendency to confine identities and to deny the history and the evolution of human societies. Focusing on wetlands problematises this view. For Bodin, wetlands were microclimates almost impossible to change. But as early as 1599, his ideas were recovered by the monarchy, which used it in a very effective way. The king stated that microclimates could be changed by human action. In this way, climate theory was used to think about change: the climate was the fruit of human actions and decisions. If society and behaviours change, then the climate would change with them. That is why, through climate theory, the environment became a tool of government. In targeting environmental change, governments were actually concerned with population and the legal order. In that sense, it is striking that the continuous drainage policy led to very important results. Even if a large part of them could have been abandoned or could have failed, French landscapes were deeply influenced and marked during these times. As a matter of fact, royal legal disposition decisively helped desiccators to complete important projects. In every case, they modified the way of life in specific areas.

Climate theory progressively gave way to liberal and economic thought as a means by which to justify draining works. Before the end of the eighteenth century, the individual interest of landowners was the most important stated reason for supporting drainage. The state no longer gave the right to few people to drain the marshes but recognised the right of each landowner to increase his own cultivated soil and thus to profit. Nevertheless, the justifications advanced by the first Bourbon remained influential for a very long time and played a key role in French culture and its relationship to its environments. Thus, the edict of 1599 was quoted throughout the nineteenth century and is still remembered by local actors. During the French Revolution and thereafter in the nineteenth century, desiccation was conceived as a means to give work to unemployed people. Even more strikingly, the legacy of climate theory can be seen in more recent colonial experience. The negative vision of marshes and their inhabitants has been intensively used by French settlers in Algeria, in particular in the Mitidja plain.[76] Promoted in the early modern period, such a negative vision has been commonly shared until only recently, enshrined within renewed concept frameworks such as modern geography and the ideology of progress.

Notes

1 Paul Warde, 'Fear of Wood Shortage and the Reality of the Woodland in Europe, c. 1450–1850', *History Workshop Journal* 62 (Autumn, 2006): 28–57; Edward Palmer Thompson, *Whigs and Hunters: The Origin of the Black Act* (London: Penguin Books, 1975).
2 Steven Lawrence Kaplan, *The Famine Plot Persuasion in Eighteenth-Century France* (Philadelphia: Transactions of the American Philosophical Society, 1982).
3 Paul Warde, *Ecology, Economy and State Formation in Early Modern Germany* (Cambridge: Cambridge University Press, 2006); Karl Appuhn, *A Forest on the Sea: Environmental Expertise in Renaissance Venice* (Baltimore, MD: John Hopkins University Press, 2009).
4 Jean Boissière, 'Population et économie du bois dans la France moderne. Contribution à l'étude des milieux forestiers entre Paris et le Morvan au dernier siècle de l'Ancien régime (vers 1685-vers 1790)' (PhD thesis, Université Paris 1, 1993); Grégory Quenet, *Versailles, une histoire naturelle* (Paris: La Découverte, 2015).
5 Jean Bodin, *The Six Books of a Common-Weale: Written by I. Bodin a Famous Lawyer, and a Man of Great Experience in Matters of State: Out of the French and Latine Copies, Done into English, by Richard Knolles* (London: G. Bishop, 1606), 571.
6 Michel Foucault, *Sécurité, territoire, population* (Paris: EHESS-Gallimard-Le Seuil, 2004).
7 Vincent Denis, *Une histoire de l'identité, France, 1715–1815* (Seyssel: Champ Vallon, 2002).
8 Michel Foucault, *Naissance de la biopolitique* (Paris: EHESS-Gallimard-Le Seuil, 2004), 46–47.
9 See the Editors' introduction to this volume.
10 Clarence Glacken, *Traces on the Rhodian Shore: Nature and Culture in Western Thought from Ancient Times to the End of the Eighteenth Century* (Berkeley: University of California Press, 1967).
11 André Miquel, *La géographie humaine du monde musulman jusqu'au milieu du XIe siècle* (Paris and La Haye: Mouton, 1975), 56–68.
12 Thomas Aquinas, *De regimine principum*; Aristotle, *Politics*, Book 7, 1327b; Frank Lestringant, 'Europe et théorie des climats dans la seconde moitié du XVIe siècle', in *Écrire le monde à la Renaissance: quinze études sur Rabelais, Postel, Bodin et la littérature géographique*, ed. Frank Lestringant (Caen: Paradigme, 1993), 206–226.
13 Marie-Dominique Couzinet and Jean-François Staszak, 'À quoi sert la "théorie des climats"? Éléments d'une histoire du déterminisme environnemental', *Corpus* 34 (1998): 5–43.
14 Salvatore Ciriacono, *Building on Waters: Venice, Holland and the Construction of the European Landscape in Early Modern Times* (Oxford and New York: Berghan Books, 2006); Petra van Dam and Milja van Tielhof, *Waterstaat in Stedenland. Het Hoogheemraaschap van Rijnland voor 1857* (Utrecht: Matrijs, 2007); Paul Slack, *The Invention of Improvement: Information & Material Progress in Seventeenth-Century England* (Oxford: Oxford University Press, 2015); Vera S. Candiani, 'The Desagüe Reconsidered: Environmental Dimensions of Class Conflict in Colonial Mexico', *Hispanic American Historical Review* 92, no. 1 (2008): 5–39; see also, Chapter 8 in this volume, Anthony E. Carlson, 'Vast Factories of Febrile Poison: Wetlands, Drainage, and the Fate of American Climates, 1750–1850'.

15 Clive Holmes, 'Drainage Projects in Elizabethan England: The European Dimension', in *Eau et développement dans l'Europe moderne*, ed. Salvatore Ciriacono (Paris: MSH, 2004), 87–102.
16 *Edit portant règlement pour l'aliénation du domaine*, Paris, P. J. Mariette, 1728 [1566], 7.
17 'Ordonnance sur le fait des eaux et forêts, Vérifiée en Parlement et Chambre des comptes le 13e aoust 1669' (Paris: P. Le Petit, 1669).
18 *Edit pour le desséchement des marais, portant commission à cet effet à un étranger*, Fontainebleau, 8 avril 1599.
19 Denis Diderot and Jean d'Alembert, eds., *Encyclopédie ou dictionnaire raisonné des sciences, des arts et des métiers*, vol. 3 (Paris: Briasson – David – Le Breton – Durand, 1753), 532–536.
20 Hippocrates, 'Airs, Waters and Places', in *Hippocrates*, vol. 1, translated by W.H.S. Jones (Cambridge, MA: Harvard University Press, 1923), 114. The Phase is now called the Rioni and flows in the western part of Georgia.
21 Ibid., 114.
22 Ibid., 114.
23 Ibid., 113.
24 Pierre Briant, ed., *Irrigation et drainage dans l'Antiquité, quanuts et canalisation souterraines en Iran, en Egypte et en Grèce* (Paris: Collège de France/Thotm Edition, 2001).
25 Vitruvius, *On Architecture* (Cambridge, MA: Harvard University Press, 2002), Book 1, 45.
26 Columella, *On Agriculture* (Cambridge, MA: Harvard University Press, 2010), Book 1, 63.
27 Philippe Leveau, 'Revisiter l'espace et le temps dans le delta du Rhône: archéologie et histoire des zones humides et des milieux deltaïques', *Bulletin Archéologique de Provence*, Supplément 2 (2003): 13–44.
28 F. Spada and S. Passigli, 'Paesaggio, risorse e percezione dell'ambiente nelle zone umide: esempi dall'area romana medievale (secoli X–XV)', in *Les zones humides méditerranéennes hier et aujourd'hui / Le zone umide mediterranee ieri e oggi*, ed. Magalie Franchomme, Christine Labeur, Daria Quatrida and Remy Simonetti (Padua: Padova University Press, 2014), 85–102.
29 Leon Battista Alberti, *On the Art of Building in Ten Books* (Cambridge, MA: The MIT Press, 1991).
30 Ibid., 13.
31 Denis Cosgrove, 'Platonism and Practicality: Hydrology, Engineering and Landscape in Sixteenth-Century Venice', in *Water Engineering and Landscape: Water Control and Landscape Transformation in the Modern Period*, ed. Denis Cosgrove and Geoff Petts (London and New York: Bellhaven Press, 1990), 35–53.
32 *Genesis*, 1, 9; see also, Andries Vierlingh, *Tractaet van Dyckagie*, ed. J. de Hullu and A.G. Verhoeven (The Hague: Martinus Nijhoff, 1920). Vierlingh's treatise was originally written during the 1570s.
33 Sara Miglietti, 'Amitié, harmonie et paix politique chez Aristote et Jean Bodin', *Astérion* [Online], 7 (2010), online since 1 September 2010, consulted 19 July 2016. URL: http://asterion.revues.org/1660.
34 Jean Bodin, *Six Books of a Common-Weale*, Book 5, Chapter 1.
35 Ibid., 547.
36 Ibid., 565.
37 Montesquieu, *De l'Esprit des Loix, ou du rapport que les loix doivent avoir avec la constitution de chaque gouvernement, mœurs, climat, religion, commerce, etc. (sic); à quoi l'auteur a ajouté des recherches sur les lois romaines touchant les successions, sur les lois françaises et sur les lois féodales*, 2 vols. (Genève: Barrillot & Fils, s.d. [1748]).
38 D'Alembert, 'Climat', in *Encyclopédie ou dictionnaire raisonné*, vol. 3, 534.
39 Montesquieu, *The Spirit of the Laws* (Cambridge: Cambridge University Press, 1989), Book 18, Chapter 7, 289.
40 Ibid., Book 18, Chapter 19, 294.
41 Ibid., Book 18, Chapter 10, 290.
42 Mario Pinna, *La teoria dei climi. Una falsa dottrina che non muta da ippocrate a Hegel* (Rome: Società geografica italiana, 1988), 99ff.

43 Ibid., 117; Eileen E. Power, 'Pierre Dubois and the Domination of France', in *The Social and Political Ideas of Some Great Medieval Thinkers*, ed. F.J.C. Hearnshaw (London: Harrap & Company, 1923), 139–166.
44 Mario Pinna, *La teoria dei climi*, 117; Serge Lusignan, *La construction d'une identité universitaire en France* (Paris: Publications de la Sorbonne, 1999).
45 Jean de Jandun, 'Tractatus de Laudibus Parisius', in *Paris et ses historiens aux XVIe et XVe siècles. Documents et écrits originaux*, ed. Le Roux de Lincy and L.M. Tisserand (Paris: Imprimerie impériale, 1867), 60.
46 Mathieu Arnoux, *Le temps des laboureurs. Travail, ordre social et croissance en Europe (XIe–XIVe siècle)* (Paris: Albin Michel, 2012).
47 Hilton Root, *Peasants and King in Burgundy: Agrarian Foundations of French Absolutism* (Berkeley: University of California Press, 1987), Chapter 1.
48 Jean-Luc Pinol, ed., *Histoire de l'Europe urbaine. I. De l'Antiquité au XVIIIe siècle* (Paris: Le Seuil, 2003), 614.
49 Pierre Charron, *De la Sagesse*, vol. 3 (Paris: Fayard, 1986), 578.
50 Joël Cornette, *Le roi de Guerre* (Paris: Payot, 2000).
51 Cornette, *Le roi de Guerre*, 239ff.; Gérard Sabatier, *Versailles ou la figure du roi* (Paris: Fayard, 1999).
52 Kaplan, *The Famine Plot*, 52.
53 Steven Laurence Kaplan, *Le Pain, le Peuple et le Roi. La bataille du libéralisme sous Louis XV* (Paris: Perrin, 1986).
54 '[. . .] il y a grande quantité de palus et marais inondez et entrepris d'eau, et presque inutiles, et de peu de profit, qui tiennent beaucoup de pays comme désert et inhabité, et incommodent les habitants voisins, tant à cause de leurs mauvaises vapeurs et exhalations, que de ce qu'ils rendent les passages fort difficiles et dangereux: lesquels palus et maraiz estans desseichez, serviront partie en labour et partie en prairies de pasturages', *Edit pour le desséchement des marais, portant commission à cet effet à un étranger*, Fontainebleau, 1599.
55 'Comme le feu roy Henry le Grand d'heureuse memoire, nostre très honoré Seigneur et Pere, avoit bien reconneu que ce que la terre produit dans l'éstenduë des lieux de son obéissance, (si utiles à ses sujets, & nécessaires à ses voisins) lui servoit plus que ne sont les Indes aux Princes qui s'en prevalent', in *Recueil des édits, declarations, arrêts et règlemens, concernant le deseichemens des marais, S. Boé, Bordeaux* (Paris, 1639), 1681.
56 Marc Lescarbot, *Histoire de la Nouvelle France* (Paris: chez Jean Milot, 1609), 624–625; Pierre Biard, 'Relation de la nouvelle France [. . .] faite par le père Pierre Biard [1616]', in *Relations des Jésuites dans la Nouvelle France*, vol. 1 (Québec: Augustin Coté, 1858), 1–76. Both cited in Jean-Baptiste Fressoz and Fabien Locher, 'L'agir humain sur le climat et la naissance de la climatologie historique, XVIIe–XVIIIe siècles', *Revue d'Histoire Moderne et Contemporaine* 1 (2015): 48–78.
57 *Edit du roy, pour le desséchement de étangs, palus et marais u bas Languedoc, depuis la ville de Beaucaire jusqu'à celle d'Aiguemortes et à l'étang de Perols* (F. Léonard. Paris, 1702).
58 André Bourde, *Agronomie et agronomes en France au XVIIIe siècle* (Paris: SEVPEN, 1967), 1077ff., 1447–1464.
59 Very long-term leases, generally 99 years.
60 *Déclaration du roi qui permet à tous seigneurs et propriétaires de marais palus et terres inondées d'en faire le desséchemens vérification préalablement faite de l'état et consistance desdits terrains* (Versailles: Imprimerie royale, 1764) ('qu'il soit libre et permis, comme nous avons permis et permettons à tous seigneurs et propriétaires de marais, palus et terres inondées, ensemble à tous ceux qui en ont ci devant pris et prendront ci-après par baux emphytéotiques ou à perpétuité à droit de champart, de faire les desséchements desdits marais, palus et terres inondées').
61 André Bourde, *Agronomie et agronome*, 1501–1502.
62 See Chapter 8 in this volume, Anthony E. Carlson, 'Vast Factories of Febrile Poison: Wetlands, Drainage, and the Fate of American Climates, 1750–1850'.
63 Raphaël Morera, 'Dutch Capital and the Draining of French Wetlands in Arles and Petit Poitou: Environmental Change within an International Network in the Seventeenth

Century', *International Review of Social History* 55 [Special issue supplement 18] (2010); Raphaël Morera, *L'assèchement des marais en France au XVIIe siècle* (Rennes: Presses universitaires de Rennes, 2011), 128–144.

64 *Déclaration du roi qui permet à tous seigneurs et propriétaires de marais palus et terres inondées d'en faire le desséchemens vérification préalablement faite de l'état et consistance desdits terrains* (Paris: Imprimerie royale, 1764) ('qu'il soit libre et permis, comme nous avons permis et permettons à tous seigneurs et propriétaires de marais, palus et terres inondées, ensemble à tous ceux qui en ont ci devant pris et prendront ci-après par baux emphytéotiques ou à perpétuité à droit de champart, de faire les desséchements desdits marais, palus et terres inondées').

65 *Lettres patentes portant don des Moëres en faveur de la dame marquise de Maisons et du sieur marquis de Canillac* (Paris: veuve Saugrain et P. Prault, 1716); *Arrêt du Conseil d'Etat et lettres patentes portant don des Moeres en faveur du sieur Compte d'Hérouville de Claye, lieutenant général des armées de Sa Majesté* (Paris: Prault, 1765).

66 Archives nationales, Paris (France), (hereafter AN), G7 / 1673.

67 Jean-Laurent Rosenthal, *Fruits of Revolution: Property Rights, Litigation and French Agriculture 1700–1860* (Cambridge: Cambridge University Press, 1992).

68 Root, *Peasants and King in Burgundy*, 22–45.

69 AN, G7 1673: 'le desséchement causera encore à plus de quinze lieues à la ronde une sécheresse à tout perdre: personne n'ignore que des marais d'une aussi vaste étendue étant dans ces cantons la cause principale des pluyes, elles ne peuvent que cesser ou y devenir très rares, dès qu'on les aura desséchez'.

70 Raphaël Morera, 'Mise en valeur des zones humides et associations de gestion. Naissance et affirmation de nouveaux pouvoirs territoriaux (France, XVIe-XVIIIe siècles)', *Siècles* [online] 42 (2015), online since 17 March 2016, consulted 6 September 2016. URL: http://siecles.revues.org/2946.

71 Yannis Suire, *Le marais Poitevin. Une écohistoire du XVIe siècle à l'aube du XXe siècle* (La Roche sur Yon: Centre vendéen de recherches historiques, 2006), 177–227.

72 Raphaël Morera, 'La gestion de l'eau dans le Calaisis, les pouvoirs civils et militaires face aux populations locales, fin du XVIIe–début du XVIIIe siècle', *Aménagement et environnement*, ed. Patrick Fournier and Geneviève Massard-Guilbaud (Rennes: Presses universitaires de Rennes, 2016), 25–40.

73 *Rapport du sieur Bernardy ingenieur, du 29 avril 1733*, Bibliothèque municipale d'Arles, Fonds Véran, 493.

74 *Déclaration du roi qui permet à tous seigneurs et propriétaires . . .* (Paris: Imprimerie royale, 1764); Archives municipales, Lourdes, E 317.

75 See, in this volume, Richard Spavin, 'Jean Bodin and the Idea of Anachorism'.

76 Nadjet Aroua, 'Les marais de la Mitidja, enjeux et moteur de la colonisation en Algérie', in *Les zones humidese méditerranéennes hier et aujourd'hui*, ed. Franchomme, Labeur, Quatrida and Simonetti (Padua: Padua University Press, 2015), 47–64.

4 Mastering north-east England's 'River of Tine'

Efforts to manage a river's flow, functions and form, 1529–c.1800

Leona Skelton

Introduction

England's a perfect world!
Has Indies too!
Correct your maps:
Newcastle is Peru.

William Ellis, *News From Newcastle* (London, 1651)

Early modern rivers have been depicted as essential liquid highways for trade, as useful waste-disposal receptacles into which all manner of filth was dumped and sometimes, when naturally they burst their banks to use their flood plains, as violent and frightening agents of economically and socially devastating flood events.[1] But these powerful seaward forces are rarely discussed explicitly as natural systems interacting with wind, tidal fluctuations, rain and ice or as the objects of sensitive environmental management underpinned by relatively extensive riverine knowledge. Social, cultural and economic histories of rivers have tended to reduce them to little more than stages or backdrops against which human histories have been played out.[2] Few pay attention to rivers' natural functions such as tides, their independent agency to use their flood plains or to divert their courses or their integral and long-established contributions to wider ecosystems, landscape and hydrological cycles. The 'conservators' who regulated riparian development and the disposal of ballast and solid waste into rivers, who understood to a high degree, out of necessity, the natural rhythms of their local rivers and the impact of wind, heavy rainfall, tidal fluctuations and ice on a river's natural functions are very often missing from narratives of early modern rivers.[3] This leads readers to assume far too readily that this was an age of environmental ignorance. It was not.

It is well established that north-east England's River Tyne was a bustling hive of industry throughout the early modern period, receiving and dispatching thousands of ships, most notably the plethora of colliers which transported crucial so-called 'sea coal' from Newcastle down the east coast to London.[4] In 1932, John Nef conjured up a well-informed, albeit imaginative, picture of the riverscape as it changed dramatically in the century following Elizabeth I's coronation in 1558:

> Picture the mouth of the muddy, narrow River Tyne, jammed with four or five hundred keels and two or three hundred ships . . . think of the hilly slopes to the north and south covered with hundreds of small carts and wagons, leaving behind them trails of black refuse on the green countryside; and then think of a time when this same countryside was at rest . . . when the only evidence of the coal industry was a few pits at the water's edge . . . In this comparison you have . . . a view of the change wrought around the town of Newcastle in the century following the accession of Elizabeth [i.e. 1558].[5]

Much attention has been paid by historians to the social, economic and political impact of dramatic industrialisation which centred around and relied very heavily on the natural functions of the navigable River Tyne.[6] This chapter places the river water, channel and bed and two-way socio-environmental interactions with the river at its heart.

In 1529, the English Crown confirmed the mayor and burgesses of Newcastle as conservators of the Tyne's tidal estuary by an Act of Parliament (21 Hen. VIII, c.18). This decision was made at the height of the Reformation, when King Henry VIII favoured municipal over clerical government. Previously, Newcastle Corporation fought the Bishops of Durham for the conservatorship throughout the fifteenth century. The 1529 act gave the corporation authority on behalf of the Crown to pull down all weirs, gores and engines in the river Tyne between Sparrow-Hawk and Hedwin Streams, which marked the Tyne's estuary, and to exercise a monopoly over all ballast and shipping tolls. Subsequently, several successive acts reconfirmed their conservancy and, in 1613, the corporation appointed twelve River Jurors to regulate the river by enforcing twenty-one bylaws at a weekly court. The eight-mile route upriver was angular and complex, often requiring a pilot of Newcastle Trinity House, established in 1505, to deliver ships safely to the port at Newcastle via particular pathways known as 'roadsteads', in return for a fee, past shoals, rock projections, sunken and wrecked ships awaiting removal and sand bars.[7] One might assume that the arrival of steam ships motivated a heightened desire to maintain a wide river, but sail ships required even greater room for error than their steam-powered descendants. Therefore, early modern motivations to maintain a deep and wide channel were strong. Newcastle Corporation and its River Jurors were principally concerned about protecting their income from shipping and ballast tolls by ensuring that the Tyne did not silt up. However, in the course of achieving this, they engaged with climatic influences on the river's natural functions in highly sophisticated ways to achieve at least some mastery over the river's flow, form and functions, increasingly from 1529 to the dawn of the nineteenth century, harnessing as best they could the adverse effects of wind, heavy rain, tides and ice.

While the River Jurors worked hard to regulate the Tyne estuary in ways which happened to be beneficial to the river's environmental health, they were not motivated by environmentalism in the modern sense of the word. They were driven, first and foremost, to protect their monopolistic income from shipping and ballast tolls. They protected the Tyne estuary from harm – as they perceived it – for their own sake and in the economic interests of the city they governed. In this sense, their environmental regulation was distinctly different from today's, which results from motivation to protect a river from perceived harm for its own sake. Those who managed early modern rivers did not leave behind detailed letters explaining their perception of and relationship with rivers. As Barbara Hanawalt and Lisa Kiser observe, 'for many writing in these early periods, "nature" was arguably not even a discursive category; it simply went without saying'.[8] In 1995, Richard White asked us to 'look for the natural in the dams and the unnatural in the salmon'.[9] Perhaps Richard White's concept of the total entanglement between the human and the natural would raise few eyebrows in a seventeenth-century coffee house; he would merely be stating a widely acknowledged fact. Joel Kaye argued that while between 1250 and 1350 the 'phrase and the concept' of a balance of nature was unknown, 'the sense conveyed by the phrase balance of nature was very much alive and active in scientific speculation'.[10] Underlining Kaye's argument, Fiona Watson found 'little evidence' to suggest that contemporaries 'regarded timber as a resource to be plundered at will'.[11] While she warns historians to be 'very wary of transporting such [modern] values to the early modern world', she nevertheless emphasises that early modern people, faced with having to 'stay and face the consequences of their inefficiency', were highly driven to work with, rather than to plunder, their environment.[12]

Richard Grove pioneered the idea that environmental conservation attitudes and values developed long before the environmentalism of the 1960s, rightly asserting that 'the origins and early history of contemporary western environmental concern and concomitant attempts at conservationist intervention lie far back in time'.[13] However, his argument that 'the seeds of modern conservationism developed as an integral part of the European encounter with the tropics and with local classifications and interpretations of the natural world and its symbolism' is less convincing in terms of its exclusion of pre-1600 progressive environmental attitudes and values and home-grown conservation and stewardship of natural resources.[14] This chapter demonstrates that the sort of acute awareness of natural systems and resources which, as Grove argues, developed in seventeenth-century European colonialists' minds in small-scale environments abroad (where Europeans were forced necessarily to consider carefully their extraction of natural resources and their use of natural systems) was also developing at home in the context of European environments.[15] Newcastle Corporation developed a keen awareness of the need to preserve and maintain the navigable river environment by understanding its functions in order to sustain their river economy. The Tyne's conservators were well aware of the need to protect the river water from the deposition of large volumes of ballast and solid waste in order to preserve both its depth for large ships and its efficient and powerful speed of flow, on which they so heavily relied throughout the early modern period. At this time, coal doubled in price for every two miles it was carried over land, and shipping was the prime method of transport.[16] As Christopher Smout highlights, throughout the seventeenth and eighteenth centuries, people living on Scotland's eastern coast preferred to import the abundant pine of Ryfylke and Sunnhordland from Norway, which was only about three days' sail from the Firth of Forth, because it involved much lower transport costs than hauling wood extracted from the Scottish Highlands over land.[17]

The Tyne's conservators gained and exercised their environmental knowledge through their physical experience of working with the river rather than as a result of reading natural philosophy. As Peter Coates observes of a later period,

> instead of trying to reconstruct a process of trickle down, or resort to the dubious argument that new ideas impose themselves through their compelling nature, it is more cogent to attribute the estrangement of industrial workers from nature to the direct daily experience of the brutal factory regime where the hegemony of the machine confronted them at every turn in the form of furnace, forge, crane, pump and water-wheel.[18]

Thus, Coates links the machines which people used in their daily working lives to their conceptualisation of how the universe might work as a machine too, explaining that they did not need to read philosophical works in order to develop such understandings. The River Jurors' understanding of the Tyne estuary was born out of the pragmatic necessity to manage and learn from the river itself in practical, physical and tangible ways rather than having been shaped by changes in natural philosophy. The case study presented in what follows demonstrates the need to look outside intellectual history for changes in environmental attitudes in the early modern period. The Tyne's conservators understood the river's natural functions in quite substantial depth, and they respected the river's powerful ability to exert itself naturally, sometimes in spite of and to the detriment of their own considered attempts to counteract such natural expressions. The river court minute books confirm beyond doubt that the Tyne's River Court Jurors did not abandon the management of the river to divine providence. Rather, they felt it was worthwhile amassing riverine knowledge, intervening to ensure the maintenance of the river's navigability and regulating riparian development and the disposal of solid waste. In short, the Jurors felt at least somewhat in control of the river's natural environment.[19]

Martin Schmid observes that 'we speak of "regulated" or "systematically trained" rivers from the early 19th century, but many rivers in Europe were by no means unregulated before that time'.[20] Schmid explains that despite substantial differences between 'the regulations of the 16th century and the work of a Tulla on the Rhine in the 19th century', describing early modern European rivers as '"largely unregulated" or even "natural" rivers' is inaccurate.[21] The scale of early modern industrial development was significant, and the Tyne's conservators were explicitly aware of its potential to damage the river's functions as a navigable river. Pierre-Claude Reynard highlights how 'early modern concerns about industrial effluents were not marginal or nascent, haphazard or weak', but rather they were 'full-fledged and central to the preoccupations of those who thought about a liveable environment and acted to defend it'.[22] When analysed in the context of contemporary knowledge and understanding rather than through modern-day eyes, the early modern period appears as an age of sophisticated and increasing environmental sensitivity, when substantial efforts were made to work with a river's natural functions and to protect those functions from the impact of unlimited riparian development and from the disposal of ballast and solid waste. As underlined in the introduction to this volume, when considering attitudes and values in relation to natural resources and the environment, it is very important to separate the early modern period from the subsequent, contextually different nineteenth century.

Ronald Zupko and Robert Laures warn that it is misleading to perceive 'environmental awareness' as a completely modern movement, 'arising out of the tumult of a half-century of war and depression like some Venus given birth in the crashing surf of a Mediterranean shore'.[23] Sixteenth-century concerns were, of course, born out of medieval concerns which had developed in similar veins. Richard Hoffmann 'seeks to encourage medievalists to think about the interactions between medieval society and its natural environment and to explore the ecological connections which shaped those changes', and he is keen to restore 'lost agency to natural systems which some, though not all, medievalists have tended to describe passively as objects of human action'.[24] In his wide-ranging environmental history of the early modern world, John Richards observes, 'as the term *early modern* suggests', the 'long-term trends that accelerated in this period deeply influenced massive and growing human-induced environmental change' after 1800.[25] Modern environmental historians can also appreciate the long-term foundations of their case studies. Sara Pritchard argues that environmentally progressive action predates the environmentalism of the 1960s substantially, noting,

> [L]ate-eighteenth- and early-nineteenth-century [French] laws tended to promote continuity of use – a sort of ecological as well as economic status quo – over new development. In this respect, the early post-revolutionary state generally preserved a river's 'natural state'. Although all of these laws permitted property owners along a non-navigable river to continue using its waters, they also circumscribed 'private' rights.[26]

An appreciation of the Rhone's long-term environmental history was useful to Pritchard in the course of analysing her twentieth-century case study. Stefania Barca, too, emphasises the long-term historical foundation of nineteenth-century industrialism:

> Stemming from the political economy of the enlightenment project, the factory system in fact took place in a landscape that was the historical product of long-term interactions between social and natural forces. As in many other European valleys and rural communities, industrialisation was preceded by what historians call the pre-industrial system of production or proto-industry.[27]

Thereby, Barca connects the early modern to the modern rather than arbitrarily separating the epochs.

Historians of improvement argue that the later sixteenth and seventeenth centuries saw a rise in 'exploitative' or extractive practices in agriculture, as opposed to prior medieval practices of stewardship.[28] On the Tyne, while efforts were made to maintain a navigable river, there was no sustained drive to 'improve' the river by deepening, straightening or widening it dramatically or by making large-scale investments in the efficiency of export infrastructure.[29] Indeed, the Royal Commission, which was sent to Newcastle in 1855 to ascertain why so little progress had been made on river improvement, opened by explaining that 'during the last two or three centuries all the evidence would lead to the conclusion that a very general similarity has been maintained in the state of the Tyne', confirming that 'the navigable state of the river has not been much altered'.[30] They had maintained it, but they had not improved it. Referring to the River Tyne Court minutes, Ralph Gardner's *England's Grievance Rediscovered* (1655), Newcastle Corporation minutes, descriptions of the 1771 and 1815 Tyne flood events and many other documents, the chapter argues that attempts to control and manage the Tyne became increasingly well informed and sophisticated between 1529 and 1800. The river's early modern conservators were certainly in touch with their environment, and they understood its functions in substantial depth.

Ralph Gardner's *England's Grievance Rediscovered* (1655)

In 1625, Ralph Gardner was born in Northumbrian Ponteland. He became a brewer in North Shields at the age of 23 and, along with many others in North and South Shields, he disagreed with Newcastle Corporation's oligarchic monopoly over shipping and ballast tolls, markets and brewing. For Gardner, this was grossly unjust. Following his refusal to close down his illegal brewery in North Shields, which provisioned ships, Ralph Gardner was imprisoned and charged a substantial fine. He wrote his *England's Grievance Rediscovered in Relation to the Coal Trade* during his imprisonment, probably in 1653. The day before Gardner's petition was scheduled to be heard in Parliament, in December 1653, Oliver Cromwell dissolved the Long Parliament. Consequently, the river's management remained in the hands of Newcastle Corporation until 1850, when the Tyne Improvement Commission, representing all of the riverside communities, took over the conservancy by Act of Parliament. Two years after the petition's rejection, in 1655, Ralph Gardner published his unsuccessful petition to spread his argument more widely. The document is a passionate criticism of Newcastle Corporation's oligarchic control of shipping and ballast tolls and the consequently unnecessary journey of eight miles from the mouth of the river to the Port of Newcastle, which resulted in the wrecking of many ships and the exclusion of the people of North and South Shields, situated at the mouth of the river, from establishing a port and prospering financially themselves.

As part of his argument, Gardner explains that although the men of Newcastle Corporation had been appointed as conservators of the river by the Crown, they were nevertheless unable to control the river's climate, going as far as to call them 'destroyers of the famous river of Tine'.[31] He elaborated,

> If these men could command the wind, and seas, not to rage and swell but to be hushed into a calme and the river kept from friezing until they sent down help from Newcastle their reply might be admitted but since the wind, sea and ice are not controllable by their charter, what abominable tyranny what savage inhumanity is it to deny ships in distresses.[32]

In this statement, Gardner emphasised the nominal nature of the Mayor and Aldermen's appointment as conservators and their ultimate weakness in comparison to far more powerful climatic influences on the river. By situating the port at Newcastle instead of Shields, he implied, they were audaciously imposing their own illogical design on the river, which was motivated by economic profit rather than the need to work within the river's natural characteristics. Rather than allowing the river's naturally difficult navigational passage from Shields to Newcastle to dissuade them from locating the port at Newcastle, they insisted, illogically, on forcing ships up a river which was full of natural impediments to navigation. Alternatively, he argued, by holding the port at the mouth of the river, they could avoid these natural and ultimately insurmountable problems. Gardner suggested that the Commissioners of Sewers of Northumberland and Durham would, perhaps, make more suitable conservators for a nationally strategic river, from which crucial coal was exported to London.[33] He was trying to warn Cromwell that the river was in danger of being choked up with ballast, noting that some 'three thousand tuns of ballast have fallen into the river in one nights time', which Newcastle Corporation was contractually bound to prevent.[34] Consequently, he elaborated, within the last twenty years, 'where twenty ships of a certain burden could have rid afloat in most road steads in the river at low water mark now not above four ships can ride afloat'.[35] Quoting a London merchant, Thomas Horth, who had known the river for twenty-five years, Gardner explained,

> [Twenty years ago] ships of the burden of two hundred tuns a piece could have rid afloat in most road steads in that river and now not above four or five, at a low water mark by reason they have so little ground that it is so over full and hilly with the ballast that the winds and rains every time doth wash and blow great quantities off into the river and that in one night the shoar called Bill Ballast key brake down at least three thousand tun of ballast, sand and gravel and stones fell down into the river and they never knew any taken up.[36]

Similarly, in February 1633, the Master of Newcastle Trinity House expressed similar concerns about the river becoming choked up with ballast, noting in an official declaration of Newcastle Trinity House that if

> there be not some speedie order taken that the saide staithes and keyes maye be so sufficiently built and kept in so good reparations that no quantitie

> of ballast or rubbishe doe falle from anye of the saide keyes or staithes . . . that the saide river of tyne will in shorte time be so choked that no ship of anye more than ordinarie burthen will be able to come up to the towne of Newcastle as usually heartofore they have done.[37]

Gardner argued that Newcastle's mayor and aldermen had failed to manage the river successfully, and he pleaded with Oliver Cromwell to reconsider their monopoly.

Gardner also complained that the corporation 'do hinder a trade all the winter season by reason neither ships, nor boats, can pass up the river which is often frozen below the ballast shoars, called the Bill Point and half way down the river it never freezeth lower'.[38] As the river froze above Bill Point, around half way between the river mouth and Newcastle, he argued, it would be better to work around this climatic tendency and situate the port at the mouth, thus obviating the entrapment of ships by the perennial freezing of the river from Bill Point up to the Port of Newcastle. He wrote,

> That all provisions brought in by sea are compelled up to Newcastle and there ingrossed into the free mens hands; people often going to market have lost their lives, and many starved to death in the two counties which cannot get to Newcastle market, in the winter season by reason of the great storms of snows, and the river frozen and no market allowed for the counties relief at Shields, where many thousand of passengers, sea men and inhabitants are being twelve miles from any market in the same county.[39]

Gardner passionately articulated the perceived injustice of forcing the people living at the mouth of the river to travel to Newcastle market to acquire their provisions, but his argument is not confined to practical expediency alone. Rather, he is using these pragmatic arguments ostensibly to drive forward his political and economic ambition. Newcastle Corporation were certainly fiscally motivated to maintain a monopolistic port at Newcastle, and Gardner, too, was financially motivated to break that monopoly to allow the people of North and South Shields to prosper by establishing their own ports. The climate forms a central part of Gardner's argument, but only because it enabled him to criticise Newcastle Corporation's ineffectual management of the river, providing a neat fit with his petition. Ultimately the natural and environmental reasons for relocating the port are underpinning a far more dominant economic and political argument.

This petition was born out of centuries of often highly charged and very long-established political arguments between Newcastle Corporation and the people of North and South Shields, dating back to arguments between Newcastle's mayor and aldermen and Tynemouth Priory in the twelfth century. Ralph Gardner was correct to assert that Newcastle Corporation was incapable of controlling the River Tyne's climate. However, the records kept by Newcastle Corporation prove beyond doubt that the river's official conservators, nevertheless, had developed a quite sophisticated understanding of the river's natural processes and functions and that they made substantial efforts to work with those functions and to protect them from both riparian overdevelopment and the deposition of ballast and solid waste.

Newcastle Corporation and its Tyne River Court

In most early modern English towns which were governed by a corporation on behalf of the Crown, river offences were regulated by one borough court held by the mayor and aldermen alongside many other aspects of urban life. However, very unusually in Newcastle, from 1613 until 1834, the river was regulated by a dedicated weekly Tyne River Court of twelve River Jurors appointed by Newcastle Corporation to enforce twenty-one river bylaws. A Water Bailiff was paid to inspect the river daily and report any offenders. The minutes of this weekly river court and the detailed Tyne-specific bylaws it enforced are not analysed here to ascertain prosecution rates or to deepen our understanding of urban life in Newcastle but rather because they contain very insightful details about how the river's conservators perceived the river's natural functions, how they worked with the river's natural expressions or tried to master them, and the conservators' level of understanding in relation to the river as a natural system. In addition to the river court's regulatory activities, the River Jurors worked closely with Newcastle Trinity house to conduct surveys of the river and to investigate particular problems along the channel, particularly the high sand bar, known as Tinmouth Barr, extending between North and South Shields. For example, in June 1736, the Master of Newcastle Trinity House, Sir Challoner Admiral Ogle, the elder brethren and a 'younger officer' had coffee between nine and ten o'clock and 'went down in the River Jury Barge, dined at the Low Lights [at North Shields] and between 3 & 4 went to sound the Barr had very little wind some time calm and very smooth sea had noe less than 9 ½foot at the neep tide the best off the bar'.[40]

The sixteenth-century corporation used some of the income from shipping and ballast tolls to maintain the estuary. In October 1574, it paid 4d to the Bellman for his work 'goinge 2 times aboute the towne, for charging the commons to sende downe the rever for helping to git up the shippe that is sonke at Hawkes Nest'.[41] Similarly, a record was made in May 1591, when 8d was paid to John Belman for 'going aboute to warn the towne 2 times to helpe wey a ship which was over throwne'.[42] In January 1593, Newcastle Corporation paid William Graie 2s 6d per week for effectively inspecting the river, 'looking for casting ballist into the river or other rubbish eyther above the bridge or below or in Gateshide'.[43] Although William Graie was not a sworn and elected official, he was a paid civic employee responsible for reporting those who cast ballast into the river. From 1529 to 1613, Newcastle Corporation maintained and inspected the river using individuals employed on an *ad hoc* basis, seeking out those who threw rubbish or ballast into the river and presenting and fining them at the weekly Town Court. They also paid various watermen on an irregular basis to lift sand, gravel, rocks and rubbish from the river using shovels and keel boats. During the period 1529 to 1613, Newcastle Corporation's maintenance of the estuary was characterised by informality, irregularity and reactive efforts.

After 29 January 1613, when twenty-one Tyne-specific bylaws were issued by the Trinity House in London, to be enforced by Newcastle Corporation, *ad hoc*, reactive regulation gave way to a much more proactive, centralised and regular system. The motivation and efforts to regulate the Tyne's estuary grew concomitantly with and reflected the increasing scale of industry and trade which threatened to damage it. The bylaws were comprehensive, designed to regulate many different aspects of industrial development and waste disposal on the river and riverbanks. One of the bylaws stipulated that riparian saltpan owners 'doe within six months build up their wharfs and keyes sufficiently above a full sea mark in height of the water' so that 'neither coals nor rubbish do fall into the river' and to 'carry away their pan rubbish every forty days'.[44] Others forbade using ballast to dam and back wharfs and quays 'in all parts of the river', casting ballast on wharfs below high water mark or casting ballast at North or South Shields.[45] Another instructed the corporation to inflict 'strict and severe punishment' upon masters of ships or keels who cast ballast into the river.[46] Extending the jurisdiction of Newcastle Corporation to the Tyne's tributary rivers too, they expressly forbade the construction of 'wyers, dams, or other stoppage, or casting of ballast in or near the said river or creeks', and the movement of ships during the 'night tyde' was banned outright.[47] Waste disposal high above the river in Newcastle's streets was regulated too. One bylaw ordered 'that strangers shall be appointed every week to cleanse the streets in Newcastle of their ashes and other rubbish, to prevent the rain from washing the same into the river through Loadbourn', which demonstrates a clear understanding of the connected and consequential flows from private to main open sewers, to tributary rivers and eventually into the main river.[48] Another bylaw ensured 'that all the gates on the town key be locked up every night, except one or two to stand open for the masters and seamen to go too and fro to their shipps, which will prevent servants casting ashes and other rubbish into the river' and that the gates should be watched throughout the night.[49] And another bylaw ordered that 'some trusty truly substantial men, burgesses of Newcastle, be appointed to view the river every week, and to make oath for the abuses and wrongs done unto the same'; they were to be truly objective, possessing no coals, mines nor ballast shores.[50]

According to Arthur Elliot, an eighty-three year old Newcastle keelman describing the River Jurors in 1679, 'most of them . . . [were] seamen'.[51] Ralph Tailor, a River Juror in various years after 1649, had trained as a Scrivener and by 1649 was a successful notary public; in such a capacity, he would have enjoyed the respect of Newcastle's inhabitants as a member of the middling sort, but he never served on Newcastle Corporation.[52] The Tyne River Court convened every Monday morning from January 1613 until 1834. Throughout the seventeenth century, the annually elected Water Bailiff and River Jurors inspected the river, attended particular sites to assess applications and encroachments in depth, enforced the river bylaws and presented and fined offenders at their weekly court. The River Jurors dealt with most of the applications to erect projections into the river such as wharves, jetties and galleries or weirs, but Newcastle Corporation still dealt with some river applications. However, by the eighteenth century, the latter task had been delegated to the River Jurors in almost every case. A delegation or committee of between three and twelve jurors visited the site in each case and provided a detailed report to the next court, which ultimately decided whether to grant permission for the works to commence.

Stage masters of several official ballast shores unloaded ballast from ships' holds, charging a fee payable to Newcastle Corporation. Therefore, ship masters were financially motivated to dump their ballast directly into the river. In addition, ballast heaps stored below the high water mark were vulnerable to being washed into the river, and riparian residents and business owners were ordered to fence their properties to 'prevent the banks from falling and washing into the river, with the great floods, flashes and raines'.[53] In November 1646, Mrs Alnei's 'ballast shore was the worst & two heaps of ballast were washed off the said shore – one 19 yards long containing 100 tons of ballast above the water & the other 17 yards long containing 200 tons at least'.[54] But not all riparian ballast heaps were unstable. In 1698, Mr Rawling had been heaping ballast onto his holding area at Heworth Shore on the Tyne's south bank 'soe long unconveyed that the grass grows thereon'.[55] Rawling should have conveyed it more frequently, but he was clearly capable of heaping his ballast so stably that grass was able to take root and flourish from its stable structure. Contemporaries feared ballast falling into the river, warning in one bylaw that ballast quays must be kept in good order; 'otherwise a hundred thousand tuns of ballast will fall into the river, to the destruction thereof'.[56] To counteract the inevitable silting up from fallen ballast, 'every winter season the poor keelmen and shewelmen' dug up and loaded into their keels large volumes of ballast and sand from the bed of the river'.[57] This was a slow, laborious and expensive process, thus making rigorous environmental regulation which could reduce such work essential.

Newcastle Corporation understood the river's changeable nature and ensured that townspeople also understood and worked with the river's low and high water marks. For example, in July 1645, Newcastle Corporation ordered 'carpenters who had their timber lying at low water mark to draw it up to high water mark on pain of forfeiting it if it was not removed'.[58] In February 1646, the common council and the river jury viewed the river to observe any damage done and 'the best method of remedying it', noting that 'if the said river be not in all things well & carefully preserved & the royalties, privileges & liberties maintained the prosperous state of the corporation must perish and decay, it being the source of the town's revenues'.[59] Here we can see how clearly Newcastle Corporation perceived the link between the river's preservation, by which they meant its capacity to receive and dispatch trade ships with a sufficiently deep channel, and their economic prosperity deriving from their main source of income in the form of shipping and ballast tolls. In February 1644, the Water Bailiff reported that Sandgate Midding and Key was 'all broken downe there much rubbish falling downe to the hurt of the River'.[60] Here, the River Jurors conceptualise the addition of solid rubbish to the river water, thereby heightening the river bed and jeopardising the river's capacity to receive and dispatch ships successfully, as 'hurting' the river, which at first might imply that they perceived the river as capable of feeling hurt, perhaps even as a fully personified historical agent. What they were really expressing was 'hurt' or damage to their economic prosperity and to their vested financial interest in the property of the river on which they relied so heavily. Nevertheless, their fierce defence of the river for economic reasons caused them to protect it environmentally too, albeit unintentionally. By protecting the river from solid, organic waste to protect the depth of the water, they unwittingly protected the water from high levels of deoxygenation, which could have endangered fish life. Moreover, by endeavouring to maintain a deep channel, they enabled the river to flow to the sea relatively unhindered. In May 1645, Thomas Fenwick was presented at the River Court because he 'hath barke on his key to the hurt of the river'.[61] In this case, the bark would have done the river no harm whatsoever in environmental terms, but the River Jurors strictly forbade it because it had the potential to heighten the river bed. They also forbade the dumping of ashes, ballast, hops and grains, which would not have posed serious threats of harm to the water quality, yet they permitted the disposal of any liquid waste, some of which did have the potential to deoxygenate the water, through a series of interconnected open and closed sewers draining the streets of Newcastle down the steep hill into the river water.

In August 1647, Newcastle Corporation permitted Mr Harris and Mr Haynes to have joint interest in a twenty-one-year lease of some glasshouses adjacent to the river at Newcastle, on condition that 'they should cast no rubbish from their furnace forward but only backward'.[62] This clause was included specifically to protect the river from the dumping of rubbish into it by stipulating that such rubbish had to be removed directly from the back of the property well away from the river frontage. In May 1736, some of the riverbank between Stella and Newburn on the south side of the river was washed into the river by 'flood and high tides', leading John Humble to request permission to secure the land 'from further damage' by building a quay. After having visited the site in question, the River Jurors granted him permission to construct a quay and lease of ninety-nine years on condition that he built it no more than six yards in breadth and 'at low water mark and not further into the river 150 yards in length before any part of the ground lying from the south side of the Low Ford at Newburn down the said river so far as 550 yds in length'.[63] In July 1739, Newcastle Corporation insisted that the parish church of St. Mary's in Gateshead on the south side of the river installed and then maintained a grate at the foot of Bottle Bank to prevent 'large quantities of dirt' from being 'carried into the River Tyne to the great nuisance and prejudice thereof'.[64] And, in September 1753, the corporation recorded 'there being a bank of rubbish in the River Tyne above bridge . . . which is very prejudicial and a nuisance to the said river, it is therefore ordered for the preservation of the river that a committee of the common council do order and direct the removal of the same as they shall think proper'.[65]

The organically powered nature of technology necessarily limited the scope of any attempts to alter the river channel, bed or course. Nevertheless, attempts were made to make the river more navigable and suitable for the loading and unloading of goods onto ships. Several Newcastle salt makers petitioned Parliament in 1655, complaining about heavy expenditure incurred by altering and preparing the channel for salt production and export originally, and subsequently spending 1,000 pounds to replace their wharfs which were destroyed by Royalists in the British Civil Wars:

> Considering the great charge, cost and pains bestowed and disbursed before that Manufacture [i.e. salt making] could be brought to perfection; as first, out of a waste and rockie piece of ground adjoyning upon the River of Tyne, in taking up the said rocks and stones, and in building wharfs and staiths along the said river, and after in placing salt pans thereupon; the removing of which rocks hath made the river thereabouts far better navigable than before, though with great charge to them.[66]

Salt and coal were big business on the Tyne's riverbanks. Maintaining a sufficiently deep and wide channel to facilitate the constant reception and dispatching of large sail ships exporting coal, salt and other goods and returning weighed down with ballast extracted from Norfolk, King's Lynn and the Wash, at Newcastle some eight miles inland was a high priority with a huge economic value.

Between 1679 and 1680, Newcastle Corporation fought the south bank's clerical owners, Durham's Dean and Chapter, which governed Durham as a County Palatinate, through Charles II's Exchequer Court to successfully prevent the construction of a large coal staith and ballast shore at Jarrow Slakes, land which they believed was not sufficiently firm to support the structures.[67] According to one witness in the case, Arthur Elliot, a Newcastle waterman of eighty-three years who could remember the river 'since infancy',

> severall of the river jury . . . were sent down by the mayor and burgesses to search view and try whether the ground . . . were firme ground and whether the same might be built without prejudice or damage to the river and to that purpose the said jury did take & carry with them . . . instruments wherewith to search the ground.[68]

In 1695, Hugh Liddle, a sixty-two-year-old waterman of Hebburn on the south bank, recalled having 'tried the said place called Jarrow slake with a long poll of fower fathome length and the s[ai]d poll sunk in the s[ai]d ground called Jarrow slake soe deep that . . . [he] could not recover the same but lost the said poll in the Jarrow Slake'.[69] The Corporation won the case; clearly, Jarrow Slake was unsuitable for development. The Tyne estuary needed to be regulated in order to protect such sites from inappropriate, financially motivated and indeed dangerous overdevelopment.

As Ralph Gardner made very clear in his *Grievance Rediscovered*, the freezing of the river above Bill Point proved very inconvenient for the purposes of navigation and trade because ships moored at Newcastle could be effectively trapped there for periods as long as a few months. By November 1817, the River Jurors were observing the formation and breaking up of the ice in order to understand it more clearly.

> The said jury delivered in a report in writing the substance of which was that they begged leave to suspend their opinion till the spring of the then coming year to give them an opportunity of witnessing the effects of the weather floods and the breaking up of the ice in winter and thereupon it was ordered that all further proceedings or the said presentment should be stayed.[70]

Here, they record their request to Newcastle Corporation to suspend decision – making in relation to nuisances and encroachments on the river until the following spring, by which time they will have had time to observe the impact of the ice on the sites in question. The River Jurors were clearly well aware that the formation and breaking up of ice had a powerful potential to influence the impact of an encroachment or a riparian nuisance on the river.

In a case submitted to the River Court in June 1771, five months before the Great Flood swept several bridges away and dramatically restructured the riverbed's topography, John Moses, esquire, of Hull, applied for permission to build up his stone wear on King's Meadows, a Tyne island which he owned. Recently, it had 'failed by means whereof the depth of water there hath been reduced and the fishery greatly prejudiced'.[71] The next month, after the Jurors had visited the island, they granted permission, having given very detailed consideration to the consequences of their decision. They recorded that the weir 'being composed of loose stones laid together is an improper one' because 'such stones for want of a proper fixture will be liable to be driven by the tides further into the said river', but a stronger weir 'would be of service to the river as well as to the fishery by giving such a check to the rapidity of the current there', and this would 'save some parts of the said island beneath from being washed into the river'.[72] Throughout the eighteenth century, financially motivated witnesses were potentially powerful, and indeed ubiquitous, arms of river bylaw enforcement. Similarly, Fiona Watson attributes the success of pre-modern Baron Courts to regulate Scottish woodland management to 'the fact that the community as a whole was actively involved in it' as foresters and as '"sufficient witnesses" in any case requiring further investigation'.[73]

The great floods of 1771 and 1815

In times of flood, Novocastrians became acutely aware of how the socio-natural systems which their river was a part of could disrupt as well as facilitate their usually productive relationships with the Tyne. Floods reminded the river's conservators of the Tyne's insurmountable natural power and confirmed their ultimate powerlessness to control the river completely. However deeply they understood the river's natural functions, and no matter how stringently they regulated human interactions with it, Newcastle Corporation could not prevent the river from using its flood plains from time to time. Mark Cioc reminds us in his 'eco-biography' of the River Rhine that floods are natural events. He explains that the word '"Flood" is a highly anthropocentric term, rooted in the human proclivity to think of a river as having a fixed length but no prescribed breadth, with the result that the floodplain is often used for farms and settlements as if it were not part of the river system'.[74] 'During high-water periods, rivers absorb the extra water much like a python digests its prey: a bulge (or swell) appears as the water passes downstream'.[75] Floods are one of the most serious and threatening ways in which a river and its interactions with the climate can impact directly on the people living and working on and around it. In 1772, Rev Isaac Farrer, Curate of Egglestone, wrote a narrative of the Great Tyne Flood which occurred in November 1771, which he called 'the most dreadful inundation that ever befell that part of the country', resulting in 'a scene of horror and devastation, too dreadful for words to express, or Humanity to behold without shuddering'.[76] Typical of many other flood accounts of the period, and very far from Cioc's modern-day conceptualisation of the flood as a naturally beneficial necessity, the Curate focuses exclusively on its adverse impact on human society, perceiving it wholly negatively.

The flood resulted in the destruction of the Tyne Bridge, on which many houses and shops had been constructed. That the Curate also described destruction further upriver suggests a connection between riparian communities both above and below the tidal reach at Hedwin Streams (near Wylam):

> But Newcastle did not alone suffer from the terrible violence of this flood: Hardly a village or cottage-house from Tyne-head, in Alston-moor, to Shields, escaped its destructive fury: The bridges at Alston, Ridley Hall, Haydon, Chollerford, and Hexham, were all carried away; the wooden bridge at Allendale was swept away entire, and discovered the next day lying across a lane near Newbrough, as exactly as if fixed there by human means. It is impossible to ascertain the prodigious number of horses, black cattle, sheep and other animals, that have perished and of corn and hay stacks, hedges, fences, implements of husbandry, and whole acres of ground, which have been swept away by the impetuosity of the torrent, whereby families who have lived in affluence and plenty are now reduced to the most abject misery and want.[77]

He goes on to describe how 'many people were taken out of their houses through the roofs &c. The shrieks of women and children, frantic with all the agonies of despair, will better be conceived than described: Dead bodies and coffins were torn out of church-yards, and the living and the dead promiscuously clashed in the torrent'.[78] The flood had changed the riverbed considerably, 'the bed of the river Tyne being entirely altered by the flood, the Master and Brethren of the Trinity-house ordered the pilots to make a survey of the new channel in order to qualify themselves to lay the buoys in the proper places, that ships may be conducted up and down with usual safety'.[79]

A few decades later, in December 1815, John Bell recalled 'severe and widespread injury occasioned by the overflowing of the river'.[80] He noted that as a result of the overflowing water, whose 'appearance was particularly awful', the 'Quayside and lower part of the Close were overflowed and almost every cellar filled with water'.[81] 'Throughout the whole of Saturday the fire engines were employed in pumping the water out of the several cellars'.[82] Damage to agriculture was also severe:

> Many horses and cows have been drowned, both in their pastures and stalls and the quantity of sheep which have been drowned and washed away is very great. . . . Very great injury has also been done to the crops of corn on all the low grounds near the river, the soil off many of the fields, being entirely washed away, and others covered with sand &c.[83]

John Bell personified the river, recalling 'the sudden rise and fury of the torrent'.[84] And he noted, 'The inhabitants of many cottages situated near the banks of the river at Newburn and Scotswood had very fortunate escapes, the water having entered their houses whilst they were in bed and they were obliged to be taken out by boats through the roofs and windows'.[85] The Tyne Bridge, which had been rebuilt after the flood of 1771, stood firm, but the bridge over the Tyne at Haydon Bridge was damaged, and 'Eals bridge across the South Tyne in the parish of Knaresdale was also carried away'.[86]

Even before these two particularly severe flood events, Newcastle's mayor, aldermen and River Jurors understood the river's considerable power to move sediment, riverbanks and water with its own agency. The floods merely sharpened their awareness of the awesome power of the natural system which flowed through their town, their lives and, perhaps most importantly for them, through their livelihoods too. These natural flood events confirmed beyond doubt that no matter how much riverine knowledge they amassed and no matter how stringently they regulated the river, they would never be able to control it completely.

Conclusion

Newcastle Corporation and its River Jurors was an oligarchic local governing institution which was driven primarily, though perhaps not exclusively, to protect its economic monopoly over shipping and ballast tolls. They managed the river estuary using a relatively high degree of sophistication when it is appreciated in the context of contemporary knowledge and understanding and the still primarily organic nature of technology. They understood the climate and the tides, and they were not afraid to attempt to manage the local environment of the estuary to protect the natural functions of *their* liquid highway on which they relied so heavily. Newcastle Corporation did not understand the chemical changes they facilitated by permitting urban sewers and riparian businesses to discharge untreated liquid waste into the river water; but they considered the consequences of each and every structural change they sanctioned to the bed and channel of the river and the many applications they received to construct projections into it in great depth and detail. Moreover, they expressly forbade the deposition of any solid waste or ballast into the river water, either directly or indirectly, something which required substantial and sustained effort to regulate. Early modern attitudes and values in relation to the environment should be appreciated as part of the process, in the long term, of developing the basic character of more recent attitudes, which were crystallised in the environmentalist movement of the 1960s and 1970s. Early modern societies created important political and legal frameworks within which they could and did purposely protect natural resources, systems and landscapes from harm, as they perceived it, before in-depth scientific experimentation confirmed exactly what could and could not damage natural systems in the long and short term. It is also important to listen to the pre-modern and early modern voices which survive intact in archival documents such as the River Tyne Court Books, explaining important socio-environmental relationships and entanglements with a natural system, on the shoulders of which current environmental attitudes at least partly stand.

Environmental regulation in the early modern Tyne estuary became increasingly regularised, more proactive and ultimately more efficient over the course of three centuries. Without the Water Bailiff, the River Jurors and their weekly River Court, the Tyne estuary would certainly have become much more problematically overdeveloped, the Tyne water would have received more oxygen-hungry organic waste and complex webs of impediments to its flow would surely have been constructed to the river's detriment. Newcastle Corporation took full ownership of the pre-modern Tyne's estuary and certainly considered it to be a resource and a facility with which to generate profit and employment. Even when they used the words 'damage', 'destruction', 'hurt' and 'spoil', they were referring to damage done to the efficiency of the river for trade, first and foremost. But their proactive regulation and the time and effort they invested into controlling and ultimately limiting the adverse impacts of human activities on the river did have a positive effect on the river's condition. Newcastle Corporation and Trinity House worked together to complete detailed surveys of the river, and they were able to navigate ships from Shields to Newcastle and back because they understood every inch of the river's channel and bed. Contemporaries clearly feared 'damage' to the river, 'hurt' done to the river and 'spoiling' the river, all their own words.

Established in Newcastle in 1613, more than a century before the Wear River Court was established in Sunderland, the Tyne River Court was an effective legal facility entrusted with increasing responsibilities to visit riparian sites, producing detailed site reports and making well-considered decisions. This court certainly cannot be described as unimportant, unnecessary or irrelevant. It represents an important development in environmental history, demonstrating a clear development from the *ad hoc* and reactive management of the sixteenth century to the regular and centralised regulation of the seventeenth century to the imaginative and more inclusive innovations of the eighteenth century which widened the community's participation in enforcing regulations. The importance and relevance of the extant, detailed Tyne River Court minute books within the context of environmental history cannot be exaggerated. The careful considerations and heartfelt concern of the Tyne River Court Jurors are heavily woven into the minute books which they created, an important testament to their close relationship with the River Tyne estuary of which they were so proud and on which they relied so heavily. Early modern people were not passive victims of nature, but nor were they active, wilful and irresponsible destroyers of it. Records of early modern environmental regulation, of which there are many more under-researched examples, are no less important than seminal environmental texts from the age of environmentalism, such as *Silent Spring*.[87]

Notes

1 P. Wright, *Life on the Tyne: Water Trades on the Lower River Tyne in the Seventeenth and Eighteenth Centuries, a Reappraisal* (Farnham: Ashgate, 2014); J. Galloway and J. Potts, 'Marine Flooding in the Thames Estuary and Tidal River c.1250–1450: Impact and Response', *Area* 40 (2007): 370–379; A. McRae, 'Fluvial Nation: Rivers, Mobility and Poetry in Early Modern England', *English Literary Renaissance* 38, no. 3 (2008): 506–534.
2 J. Schneer, *The Thames: England's River* (London: Abacus, 2005); P. Ackroyd, *Thames: Sacred River* (London: Vintage, 2008).
3 For example, Wright's *Life on the Tyne* does not even refer to the Tyne River Court minute books, which are extant between 1644–1834.
4 D. Levine and K. Wrightson, *The Making of an Industrial Society: Whickham, 1560–1765* (Oxford: Oxford University Press, 1991).
5 J. Nef, *The Rise of the British Coal Industry*, vol. 1 (London: Routledge, 1932), 29–30.
6 Wright, *Life on the Tyne*.
7 Newcastle Trinity House was established in 1505 under the supervision of London Trinity House to manage the hospitality of aged mariners and the pilots who escorted ships safely from Tynemouth to the monopolistic Port of Newcastle. It also ran a charity school, provided a base for the Seamen's Guild, it was responsible for installing and maintaining the leading lights and buoys which guided vessels into the river and safely up the channel, and it managed the weighing and retrieval of sunken ships. See G. McCombie, 'The Development of Trinity House and the Guildhall before 1700', in *Newcastle and Gateshead before 1700*, ed. D. Newton and A. Pollard (Chichester: Phillimore, 2009), 171–188.
8 B. Hanawalt and L. Kiser, 'Introduction', in *Engaging with Nature: Essays on the Natural World in Medieval and Early Modern Europe*, ed. B. Hanawalt and L. Kiser (Notre Dame, IN: University of Notre Dame Press, 2008), 2.
9 R. White, *The Organic Machine: The Remaking of the Columbia River* (New York: Hill & Wang, 1995), xi.
10 J. Kaye, 'The (Re)Balance of Nature, c.1250–1350', in *Engaging with Nature*, ed. Barbara Hanawalt and Lisa Kiser (Notre Dame, IN: University of Notre Dame Press, 2008), 86.
11 F. Watson, 'Rights and Responsibilities: Wood Management as Seen through Baron Court Records', in *Scottish Woodland History*, ed. T.C. Smout (Edinburgh: Scottish Cultural Press, 1997), 103.
12 Ibid., 104, 106.
13 R. Grove, *Green Imperialism: Colonial Expansion, Tropical Island Edens and the Origins of Environmentalism, 1600–1800* (Cambridge: Cambridge University Press, 1995), 2.
14 Ibid., 3.
15 Ibid.

16 Levine and Wrightson, *The Making of an Industrial Society.*
17 T.C. Smout, *Nature Contested: Environmental History in Scotland and Northern England since 1600* (Edinburgh: Edinburgh University Press, 2000).
18 P. Coates, *Nature: Western Attitudes since Ancient Times* (Cambridge: Polity Press, 1998), 81.
19 For detailed discussion about the belief in divine providence and the natural world, see A. Walsham, *Providence in Early Modern England* (Oxford: Oxford University Press, 1999).
20 M. Schmid, 'The Environmental History of Rivers in the Early Modern Period', in *An Environmental History of the Early Modern Period: Experiments and Perspectives*, ed. M. Knoll and R. Reith (Berlin: Lit, 2014), 19–26.
21 Ibid., 19–26.
22 P. Reynard, 'Public Order and Privilege: Eighteenth-Century French Roots of Environmental Regulation', *Technology and Culture* 43 (2002): 1–28.
23 R. Zupko and R. Laures, *Straws in the Wind: Medieval Urban Environmental Law: The Case of Northern Italy* (Oxford: Westview, 1996), 1.
24 R. Hoffmann, 'Homo et Natura, Home in Natura: Ecological Perspectives on the European Middle Ages', in *Engaging with Nature*, ed. Barbara Hanawalt and Lisa Kiser (Notre Dame, IN: University of Notre Dame Press, 2008), 13. See also R. Hoffmann, 'Elemental Resources and Aquatic Ecosystems: Medieval Europeans and their Rivers', in *A History of Water: Series II, Vol. 2: Rivers and Society: From Early Civilizations to Modern Times*, ed. T. Tvedt and R. Coopey (London: I.B. Tauris, 2010), 165–202 and R. Hoffmann and V. Winiwarter, 'Making Land and Water Meet: Cycling of Nutrients between Fields and Ponds in Pre-Modern Europe', *Agricultural History* 84, no. 3 (2010): 352–380.
25 J. Richards, *The Unending Frontier: An Environmental History of the Early Modern World* (London: University of California Press, 2003), 2.
26 S. Pritchard, *Confluence: The Nature of Technology and the Remaking of the Rhone* (London: Harvard University Press, 2011), 33.
27 S. Barca, *Enclosing Water: Nature and Political Economy in a Mediterranean Valley, 1796–1916* (Cambridge: White Horse Press, 2010), 60.
28 A. McRae, 'Husbandry Manuals and the Language of Agrarian Improvement', in *Culture and Cultivation in Early Modern England: Writing and the Land*, ed. M. Leslie and T. Raylor (Leicester: Leicester University Press, 1992), 35–62; A. McRae, *God Speed the Plough: The Representation of Agrarian England, 1500–1660* (Cambridge: Cambridge University Press, 1996).
29 Newcastle Corporation was criticised heavily for its historic lack of investment in river improvement, first by Ralph Gardner in his *England's Grievance Rediscovered* (1655). Similar criticisms continued up until the mid-nineteenth century, when the Tyne Improvement Commission was appointed by an Act of Parliament to replace the Corporation as conservators in 1850. This was followed in 1855 by a Royal Commission, which was sent by central government to ascertain why so little progress had been made in the seventeenth, eighteenth and early nineteenth centuries.
30 W. Bowles, I. K. Brunel, J. Bell, R. Fitzroy and R. Armstrong, *Report of the Commissioners Appointed to Inquire into the Present State of the River Tyne; together with the Minutes of Evidence and Appendix* (London: George Edward Eyre and William Spottiswoode, 1855), v.
31 R. Gardner, *England's Grievance Discovered in Relation to the Coal Trade* (London: R. Ibbitson and P. Stent, 1655) iv.
32 Ibid., 203.
33 Ibid., v.
34 Ibid., 70.
35 Ibid.
36 Ibid., 83.
37 Tyne and Wear Archives [hereafter TWA], GU/TH/9/1: Seamen's Guild Order Book, 1580–1664 [03/02/1633].
38 R. Gardner, *England's Grievance*, 71.

39 Ibid., 109.
40 TWA, GU/TH6/1: Trinity House Minutes, 1724–1736 [21/06/1736].
41 Anon, *Reprints of Rare Tracts and Imprints of Ancient Manuscripts Chiefly Illustrative of the History of the Northern Counties* (M. A. Richardson, 1847).
42 Ibid.
43 Ibid.
44 Ibid.
45 Ibid.
46 Ibid.
47 Ibid.
48 Ibid.
49 Ibid.
50 Ibid.
51 The National Archives [hereafter TNA], E 134/31CHAS2/EAST18: Exchequer, Office of First Fruits and Tenths, (Jan 1679–Jan 1680).
52 K. Wrightson, *Ralph Tailor's Summer: A Scrivener, His City and the Plague* (London: Yale University Press, 2011).
53 Anon, *Reprints of Rare Tracts.*
54 TWA, MD.NC/2/1: Newcastle Common Council Order Book, 1645–1650.
55 TWA, BC.RV/1/6: River Court Book, 1695–1755 [11/07/1698].
56 Anon, *Reprints of Rare Tracts.*
57 Ibid.
58 TWA, MD.NC/2/1: Newcastle Common Council Order Book, 1645–1650 [20/07/1645].
59 Ibid. [13/02/1645/46].
60 TWA, BC.RV/1/1: River Court Book, 1644–1647 [20/02/1644/45].
61 Ibid. [26/05/1645].
62 TWA, MD.NC/2/1: Newcastle Common Council Order Book, 1645–1650 [18/08/1647].
63 TWA, MD.NC/2/4: Newcastle Common Council Order Book, 1718–1743 [06/05/1736].
64 Ibid. [09/07/1739].
65 TWA, MD.NC/2/5: Newcastle Common Council Order Book, 1743–1766 [25/09/1753].
66 Anon, *The Ancient Manufacture of White Saltmaking at South and Northshields, Sunderland and Blyth, Ought to be Preserved and Encouraged, for the Ensuing Reasons Most Humbly Presented* (London, n.p., 1655).
67 TNA, E 134/31CHAS2/EAST18: Exchequer, Office of First Fruits and Tenths, (Jan 1679–Jan 1680).
68 Ibid.
69 TNA, E 134/7WM3/EAST30: Exchequer, Office of First Fruits and Tenths, (Feb 1695–Feb 1696).
70 TWA, BC.RV/1/9: River Court Book, 1817–1834 [10/11/1817].
71 TWA, BC.RV/1/7: River Court Book, 1766–1772 [17/06/1771].
72 Ibid.
73 Watson, 'Rights and Responsibilities', 104, 106.
74 M. Cioc, *The Rhine: An Eco-Biography, 1815–2000* (Seattle: University of Washington Press, 2002), 33.
75 Ibid., 35.
76 Rev. I. Farrar, *Narrative of the Great Flood in the Rivers Tyne, Tease, Wear &c on the 16th and 17th of Nov 1771: Collected from the Most Authentic Papers Yet Published* (Newcastle: T. Slack, 1772).
77 Ibid.
78 Ibid.
79 Ibid.
80 J. Bell, *An Account of the Great Flood in the River Tyne on Dec 30 1815* (Newcastle: John Bell, 1816), 3.

81 Ibid., 4.
82 Ibid.
83 Ibid., 6.
84 Ibid.
85 Ibid., 4.
86 Ibid., 6.
87 R. Carson, *Silent Spring* (London: Penguin, 2000 [1962]).

5 'Take plow and spade, build and plant and make the wasteland fruitful'

Gerrard Winstanley and the importance of labour in governing the earth[1]

Ashley Dodsworth

The work of English radical Gerrard Winstanley (1609–1676) was once said to be the result of 'a mid-life crisis of epic proportions'[2] but has since received the critical scholarly attention it deserves, with his contribution to early modern debates over the use and ownership of the earth, political organisation and authority now recognised. Yet the environmental aspects of his thought remain under-researched. This chapter contributes to recent environmentally focused Winstanley scholarship and shows that Winstanley believed that to 'govern the earth' was to govern people through the control of the natural resources upon which they depended. Labouring upon resources emerges as a significant political act for Winstanley, as the transformation of natural resources could also reshape both government and society as a whole into a new world founded on equal ownership of the earth. Winstanley's actions were informed and directed by his theory, but just as crucially, his writings were underpinned and given force by his action in physically digging the land. His political thought is therefore an embodied one, given shape and form most prominently through labour.

To demonstrate these claims, this chapter first sketches Winstanley's life and work. The subsequent section discusses his argument that private ownership of the earth and natural resources was a form of political control and the means by which a minority could maintain political power. The third section shows how Winstanley conceived of labour as a means of resistance to this system of ownership and the alternative system that labouring upon resources would create in its place. The final section explores the 'praxical' elements of Winstanley's work, his connection between theory and practice.

The role of labour within Winstanley's thought and action is implicitly acknowledged in the existing literature – all commentators refer to the role of labour, but as yet no discussion has made labour its explicit focus.[3] This chapter directly addresses labour, showing that Winstanley saw it as a means of resistance and independence, a way of reordering society and fulfilling the rights of all to their environment. This presents an alternative understanding of the means by which Winstanley believed private property in land would be overthrown and explicitly develops recent discussions of the role of praxis and embodiment in his work. Within the larger literature of politics in the early modern period, this chapter highlights a radical voice and a vision of politics in which all are capable of transforming the government of their society and emphasises the importance of controlling natural resources.

Emphasising the environmental aspects of Winstanley's thought does, however, require justification. This is particularly important, as the recent engagement of contemporary environmentalists with Winstanley's work has received critical pushback from scholars.[4] For example Ariel Hessayon has argued that 'it does [past authors] an injustice' if we 'read their texts with a present-centred perspective' and use them to legitimise a Green tradition.[5] This is because 'Winstanley and the Diggers cannot easily be accommodated within emerging Green narratives', due to the emphasis on labouring upon and consuming resources.[6] A project that seeks to examine the environmental aspect of Winstanley's thought by highlighting his understanding of labour therefore seems to run directly into the contradiction identified by Hessayon. However, it is not the aim of this chapter to try to situate Winstanley more comfortably within the Green canon; instead, it begins by recognising the centrality of labour and the use of natural resources in Winstanley's thought. Examining Winstanley's conception of labour and the role it plays within his political writing may limit 'green' readings of him. But such an examination does contribute to a clearer understanding of Winstanley's relationship to environment and government. The tensions inherent in these relationships may be of use in understanding how current environmentalists do or do not seek to use natural resources.

This analysis draws primarily on the temporal aspect of Winstanley's thought. References to his religious ideas are secondary to and in support of the claims being made about his earthly arguments. This is not to deny the religious and spiritual aspect of Winstanley's writing (indeed, to do so would be to fundamentally misrepresent it); nor does this represent an attempt to impose an artificial divide between the 'religious' and the 'political' elements of his writing. Rather, the aim is to draw attention to the more temporal aspect of his work and clarify our understanding of these arguments. In order to do this, this chapter will first introduce Winstanley and his work.

Gerrard Winstanley

In January 1648, Winstanley, then a bankrupt former cloth merchant turned visionary writer, experienced what he described as a vision from God, which said 'work together; eat bread together'.[7] Believing God had shown him a new way for mankind to live, on 1 April 1649, Winstanley and five others began to dig the common land of St George's Hill in Walton-on-Thames in Surrey in order to make this vision a reality.[8] Within a week their numbers had risen, with between twenty and thirty people working and living upon the common and the group claiming that their numbers would rise, predicting 'four or five thousand' would join them.[9]

As they continued to live and work upon the Hill, the group faced local opposition. This opposition was influenced by the group's decision to clear and burn the heath and fell the forest in order to clear space for planting. Cultivating the land for crops in this way would prevent local people from grazing their animals on the lands, which represented a serious challenge to the traditional use of these resources and the community's way of life. The group was reported to the Council of State, resulting in the two central leaders of the group, Gerrard Winstanley and William Everard, being brought before General Fairfax less than three weeks after the so-called Diggers had moved to the Hill (Fairfax found the group to be 'not worth mentioning').[10] Continuing opposition resulted in an attack on the Digger settlement, and by mid-August of 1649, Winstanley and the Diggers moved to Little Heath in Cobham.[11] As John Gurney has shown, this decision made strategic sense, as the Diggers had strong links with the local community in Cobham.[12] Here they faced less opposition from the people of Cobham, though the local gentry were just as hostile to the group's actions as they had been in Walton-on-Thames. Yet again there was an organised campaign of violence against them – Winstanley's *New-Year's Gift for the Parliament and Army* and *A Watch-Word to the City of London and the Army* describe how he was arrested for digging the common land and denied a fair trial, with a particularly vivid retelling of the theft and killing of his cows.[13] As a result of this continued opposition, by April 1650, the original Diggers were forced to permanently abandon their attempts to bring Winstanley's vision to life.[14] In August 1650, Winstanley and several other Diggers went to live and work for Lady Eleanor Douglas, an eccentric aristocrat, though this final attempt to keep the community going did not last the year. In February 1652,[15] Winstanley published *The Law of Freedom in a Platform*, his last and most accomplished work in which he set out a utopian society in which the earth and all its resources were held in common by all.[16] After this, Winstanley slipped back into society, living anonymously and surprisingly successfully – indeed, his 'remarkable rise to the ranks of the gentry is not easy to explain' – before his death in 1676.[17]

How did Winstanley interpret his vision of 'work together; eat bread together', and why did these actions provoke such hostility? Winstanley understood his vision to mean that all should work or labour upon commonly owned resources together and then share the proceeds of that labour. He took from this vision a sense of communality, of shared work and profit that bound all together and in the face of which all were equal. This was based on two central claims – that the earth and all its resources belonged to all and a subsequent opposition to the concept of private ownership of natural resources.

Reflecting the strong religious aspect of Winstanley's thought, the first of these claims was based upon scripture. Winstanley believed that God had given the world to all humanity equally, claiming 'the earth is his creation-right as well as mine' because God had created the earth to be 'a common treasury . . . for whole mankind in his branches, without respect of persons'.[18] 'Surely', he argued, 'the earth was made . . . to be a common treasury for all, not a particular treasury for some'.[19]

The shared possession and habitation of the earth reflected the Garden of Eden before the fall of mankind.[20] Winstanley linked the fall to the introduction of private property, because it represented the coming of sin, particularly the sin of covetousness, into the world. Continuing the Garden of Eden metaphor, he argued that 'this coming in of bondage is called A-dam, because this ruling and teaching power without, doth dam up the spirit of peace and liberty'.[21] Winstanley also drew heavily on the Biblical story of Jacob and Esau, in which one brother stole the birthright of the other to show how the private possession of the earth went against God's plan.

Winstanley's understanding of the introduction of private property also drew upon the concept of the 'Norman yoke'.[22] Winstanley began to draw upon this popular argument during his Digger phase as he sought to address and explain the inequality of private property.[23] The Norman yoke thesis argued that the Norman conquest of England in 1066 had overthrown the traditional rights and forms of ownership of the English people and replaced them with the Norman conception of private ownership, backed by the power of the Church and the law. Throwing off the Norman yoke would enable the English people to return to their initial state of freedom.[24] This would see 'the earth set free from all kingly bondage of lords and manors and oppressing landlords which came in by conquest'.[25] The Norman yoke had subverted God's plan for the earth, as the resources to which all were originally entitled were instead taken and exclusively owned by a minority. Winstanley therefore described private property as a sin, for those who own the earth 'live in breach of the seventh and eighth commandments *thou shalt not steal nor kill*' as the 'restraining of the earth from brethren by brethren is oppression and bondage'.[26]

This system enabled a minority to monopolise natural resources, whilst the rest were forced into dependence. This majority was therefore in the position of 'seeing and finding our selves poor, wanting food to feed upon while we labour the Earth . . . and wanting ploughs, carts, corn and such materials to plant the commons withal'.[27] Those who laboured upon resources which were owned by others were therefore classed as the 'poor and oppressed peoples of England' even if they were recompensed for their work, because they were unable to independently access their means of subsistence.[28]

In response, Winstanley developed an alternative communal model in which natural resources were held by all, and all laboured together and shared the results of their endeavours among each other. This alternative is seen both in the actions of the Diggers themselves and in Winstanley's writing, particularly *The Law of Freedom in a Platform*, in which the details of his alternative community are expanded. In this text, Winstanley suggests that the state create storehouses, where the results of shared labour on the commonly held resources would be held and then redistributed according to need: 'the earth is to be planted and the fruits reaped, and carried into Barns and storehouses by the assistance of every family: And if any man or family want Corn or any other provision, they may go to the storehouses and fetch without money'.[29] Under this system there would be two types of storehouse: one from which anyone could take either food or raw materials to work upon and a second where finished goods produced by tradespeople were taken and freely shared amongst all. This system was designed to secure the benefits of labour, particularly regarding the quality and quantity of goods available, whilst ensuring that the earth as a whole remained in common. Goods taken from the storehouse would become the exclusive, private property of those who had taken them. This emphasised that it was specifically the private ownership of natural resources that Winstanley opposed, and he insisted that 'if any do buy or sell the earth or the fruits thereof . . . they shall be both put to death as traitors'.[30] Though the death penalty was the punishment for a variety of crimes, Winstanley's insistence that buying or selling natural resources is treachery, a betrayal of the body politic as a whole, is significant.

Surveillance was central to this system. The storehouses would be staffed by 'overseers' who would ensure that all the resources harvested were taken to the storehouse and monitor those who wished to take goods away.[31] Furthermore, elected officials would monitor the people and 'all ancient men, over sixty years of age' would be 'general overseers', meaning that they were entitled to observe the behaviour of others, particularly those who did not labour upon the land or took more than they needed – such as those who 'suffer more meat to be dressed at a dinner or supper then what be spent and eaten . . . before it be spoiled'.[32] Winstanley's acceptance of authoritarianism in this instance is often said to reflect the experience of the Diggers' defeat, but it demonstrates that what he opposed about 'kingly power' was the control and ownership of the earth and the extent to which the control of natural resources extended into all aspects of life.[33] To govern the earth and all its resources was an extensive undertaking and one that must be strictly overseen.

Winstanley believed that God gave the earth to all to ensure their survival, that the disruption of this shared ownership occurred through sin and the Norman yoke, and was opposed to private ownership of resources. It is therefore clear that, as Corns, Hughes and Lowenstein have argued, 'the necessity that the land become a common treasury for all was at the heart of Winstanley's social and political vision'.[34] But how was this to be achieved? The following section argues that Winstanley believed that the government of the earth and all its natural resources was central to political control, and so it was only through their labour that the poor and dispossessed could reclaim the earth.

Private ownership as means of political control

Winstanley argued that private property as imposed by the Norman yoke was the means by which a conquering minority could control England. By parcelling out land and resources to their leaders and backing this with a new legal system conducted in a new language, Norman rulers secured their power over the English people.[35] By controlling the land and all other forms of natural resources in this way, the Norman conquerors and the rulers and aristocracy that came after them were able to control the people. This could only be achieved through control of the earth (as opposed to any other goods or form of property) as, Winstanley argued, life depended on natural resources and what can be produced from them. Furthermore, Winstanley believed that the earth and all its resources were given to all by God. The earth was unique: no other form of property was divinely distributed, highlighting the sense of perversion when it was privately owned.

Winstanley memorably described private ownership of natural resources as 'locking up the earth': property owners 'lock up the treasuries of the earth . . . and suffer it to rust and moulder while others starve for want to whom it belongs'.[36] This image of decay emphasises that it is natural resources that are the subject here, as other forms of property would not change in this way. Winstanley argued that the earth was made to be taken and laboured upon, an insistence that occurs throughout his work, and which Hessayon argues precludes him from accommodation within contemporary 'green' narratives. This extractive approach to the earth seems to reflect that of Locke (for example 'land that is left wholly to nature . . . is called, as indeed it is, waste')[37] and proto-capitalist agrarians who believed that they stood apart from nature and sought to control and transform it for the sake of profit.[38] Only through a substantial intensification of agricultural production could the Diggers' overall aims and promises be achieved. 'By Winstanley's reckoning, one-half to two-thirds of England's land was not efficiently cultivated'[39] and this could be achieved through the use of new techniques and crops. Indeed, much of the success of the Diggers' plan would have to rest on an exponential increase of agricultural production of the like encouraged by authors of 'improvement' literature, including the introduction of new techniques and vegetable crops.

This argument could be taken as a call for improvement rather than a change in government, for the better use of privately owned resources rather than the abolition of private ownership. Indeed, many of those advocating for improvement in the mid-seventeenth century argued precisely for the former approach, supporting their claims through appeals to morality and religious intention.[40] But Winstanley was not calling for general improvement – even if those who owned the land ensured that it was fully developed to fulfil the needs of all (which he believed unlikely, as he believed land owners were instead seeking their own profit), he would still be unsatisfied. As Corns, Hughes and Lowenstein argue this was improvement joined to emancipation, a point made clearer by Paul Slack, who emphasises that Winstanley believed improvement of the earth should be (indeed must be) for the benefit of 'the whole community'.[41] Winstanley called for a whole-scale change in the government and ownership of the earth, because it was the only way to ensure that all shared in the benefits of improvement.

Allowing the earth to be governed and owned by a minority prevented this labour and the real improvement for all that it could bring, with desperate consequences. First, resources were wasted whilst people starved for want of them. Winstanley developed his political action in response to some of the hardest years in English history, as bad harvests and the upheaval of war combined to make food scarce and drive prices to record highs.[42] Second, with scarce yet essential resources appropriated by the few, the dispossessed were left dependent on those who had taken them for survival, creating inequality. 'All rich men live at ease', Winstanley argued, 'feeding and clothing themselves by the labours of other men, not by their own; which is their shame and not their nobility'.[43] This imbalance within society was against God's plan, for He had created all as equals and given them equal claim to the earth. This inequality maintained an unequal distribution of resources, as those dispossessed from the resources that were rightfully theirs had to submit to working for hire, and through their hired labour, land owners could control and profit from their estates.

That ownership of the earth was the basis of the government and structure of society explains the opposition to the Digger movement – both the active opposition they faced locally and the lack of support for their cause from the state institutions, despite Winstanley's frequent addresses. Winstanley and the Diggers were striking at the heart of the existing power structure: 'And why are they so furious against us? But because we endeavour to dig up their tythes, their lawyers fees, their prisons, and all that art and trade of darkness, wherby they get money under couller of law; and to plant the pleasant fruit trees of freedom in the room of that cursed thornbush'.[44] The use of 'thornbush' as a metaphor reflects the sharp hedges that were used to shut out and physically prevent the dispossessed from accessing resources and marking out the limits of their engagement with the natural world.[45] Winstanley deliberately contrasts this with the 'pleasant' fruit tree, which not only enables access, as people can pass beneath it, but will provide fruit for all. The role of labour is crucial here – Winstanley says that they are 'dig[ging]' up the old government and aiming to 'plant' a new one. As the next section shows, this was not a metaphor.

Labour as resistance

Winstanley argued that the dispossessed should use their labour to challenge the private ownership of the earth. Labour provided a means of resistance, a way of undercutting the current system and creating an alternative. As Winstanley realised, if people were governed through the earth, then transforming that earth would transform the government and structure of society. This method of resistance was also in accordance with God's will, for labouring upon natural resources was divinely commanded. After all, once He created the earth and mankind, 'then the creating power or God gives two commands more. First to subdue the earth. And this implies plowing, digging and all kind of manuring'.[46] So if the dispossessed wished to govern themselves and fulfil the will of God, they should 'take plow and spade, build and plant'.[47]

Redirection of labour

But how can labour be used to oppose the government of the earth and create an alternative society? Provocatively, Winstanley argued that the excessive private ownership of the powerful was only possible because the dispossessed enabled it, for 'as long as they continued to work for others for hire, the slaves were complicit in their own slavery'.[48] In agreeing to labour on land and resources owned by the property owners, the labourers were implicitly denying the equal claims of all to the land as 'for one to give hire and for another to work for hire; this is to dishonour the work of creation'.[49] Indeed, Winstanley argued that there was no distinction in God's eyes between the property owners and those who worked for hire: 'the hand of the Lord shall break out upon every such hireling labourer and you shall perish with the covetous rich men that have held and yet doth hold Creation under the bondage of the curse'.[50]

Furthermore, by agreeing to work for hire on the resources they were entitled to own, the day labourers permitted the property owners to own and profit from more land than they themselves could work (and indeed enabled them to develop and use resources *without* labouring themselves). So, Winstanley argued, 'by their labours they have lifted up tyrants and tyranny and by denying to labour for hire they shall pull them down again. He that works for another, either for wages or to pay him rent, works unrighteously and still lifts up the curse'.[51] The land could only be locked up through the support and labour of the dispossessed – if they refused their support, then the system would collapse.

As Geoff Kennedy describes it, this method of opposing private ownership of resources 'hinges not so much on the throwing open of enclosures but rather on the withdrawal of wage labour', which 'fundamentally undermines the concrete basis of [enclosure]'.[52] Christopher Hill linked this plan to a strike action, a term echoed by John Gurney, who describes Winstanley's proposed action as 'the mass withdrawal of labour' and 'a wage strike'.[53] Certainly not working on privately owned resources and so 'withdrawing' their labour from the current system is key to this plan. However, what Winstanley is calling for here is not the withdrawal of labour but its *redirection*. Winstanley does not suggest that the workers refuse to work on privately owned resources until their owners give up power or redistribute the land, as implied within the notion of a strike. Rather he argues that all should refuse to work on the lands owned by others *and instead* work together upon the natural resources of the common lands for themselves. The two steps of this plan are often implied, as in Kennedy's reading, or linked together, as with Gurney's, but an explicit separation of the two is important in order to see that it is the choice of where to labour and for what purpose that is essential here. It is this active decision, this conscious redirection of labour, that enables 'the poor and dispossessed' to escape the old system of power and instead create a new alternative and so govern themselves.

Winstanley believed that the redirection of labour was the only true act of political resistance. For example, he made clear that the Diggers would

> meddle with none of your properties, but what is called commonage, till the spirit in you make you cast up your lands and goods which were got and still is kept in your hands by murder and theft; and then we shall take it from the spirit that hath conquered you and not from our swords, which is an abominable and unrighteous power and a destroyer of creation.[54]

As Gurney describes it, 'the rich would thus be reduced to the status of peasant proprietors or family producers and this would in turn provide them with greater incentives for abandoning their holdings' and join the poor who were Digging the common lands.[55] And indeed Winstanley made clear that 'we do not thereby take away other men's rights', as the Diggers laboured only on the common lands, which Winstanley held were available to all by custom (though, as McDonagh and Griffin point out, this 'was a wilful misunderstanding of the legal basis of common land').[56] There is a practicality here, in that the Digger movement was vastly outnumbered and could not have survived a physical confrontation. Yet Winstanley was also sincerely committed to non-violence as 'a matter of principle'.[57] He thought that rights secured by the sword were not truly fulfilled, as the freedom they brought was compromised and created further relations of inequality: 'freedom gotten by the sword is an established bondage to some part or other of creation. . . . Victory that is gotten by the sword is a victory that slaves get over one another'.[58] Only through labour could a new government be securely and justifiably formed.

Winstanley believed that it was only through the withdrawal and redirection of labour that the unequal government of the earth could be undermined. This was because the labour of the 'poor and dispossessed' was essential to the maintenance of the private property system and so enabled a minority to control the natural resources to which all were entitled. Only through the redirection and re-engagement of labour could the Diggers construct a new, better and more equal society.

Alternative government through labour

Labouring upon common lands would create an alternative system of government. Winstanley warned the 'lords of the manor and Norman gentry' that 'surely the power that is in them, will take the rule and government from you, and give it to people that will make better use of it'.[59] And the power that is in all is the power to labour, to extract what they need from the earth and reshape it. The Diggers would create a communal society, in which they worked the land together during the day and ate together in the evening, thus fulfilling Winstanley's vision.

Though Winstanley is the originator of the Digger movement and the one whose writings brought attention to the cause, in practice the Digger communities had no authoritarian structure. This reflected and was the result of their shared ownership of the earth. As none had a specific claim to resources, decisions must be made by all, and so the common good was promoted.[60] This equality was also reflected in religious matters, with any member of the group able to preach and speak the truth of God, as Winstanley described it. The sources of 'kingly power' in the form of kings, landlords, clergy and lawyers were thus rejected and replaced with a communion of equals who shared the burdens and benefits of labour amongst themselves and so secured the (physical and spiritual) preservation of all.

The potential scope of this alternative system can be seen in Winstanley's final text, the utopian *The Law of Freedom in a Platform*. This is the only text to fully set out Winstanley's vision of an ideal society. In this society, all natural resources would be held in common, and though there would be a variety of professions (supported by universal education), each member would contribute at harvest time. By law 'every family shall come into the field, with sufficient assistance at seed time to plough, dig and plant and at harvest time to reap the fruits of the earth'.[61] This would ensure all maintained a connection to the land and shared in the labour necessary to support society: 'it will make idle persons to become workers . . . there shall be neither beggar nor idle person'.[62] The results of this labour would be carried into storehouses for all to access. The differences between this text and his previous works have been the subject of dispute, mainly due to the difference in methods.[63] But the basis of this system of government reflects that set out in the earlier works and by the Digger movement, with communal ownership of resources designed to undercut 'kingly power'.

There are two key points to note here. The first is Winstanley's conviction that through communal ownership, direction and labour, enough for all could be produced. This conviction is based on the claim that God created the world and so made enough for all, whilst private property 'lock[s] up the treasuries of the earth . . . and suffer it to rust and moulder while others starve for want to whom it belongs'.[64] As a result, 'though there be land enough in England to maintain ten times as many people as are in it, yet some must beg of their brethren or work in hard drudgery for days wages for them, or starve or steal'.[65] The role of labour in bringing about this land of plenty can be seen through Daniel Russell's definition of labour as 'a directive principle' that turns 'something that might meet a need into something that actually does'.[66] Only reclaiming the earth and labouring upon it equally would ensure that 'the treasuries of the earth' that *could* meet the needs of all would be transformed into the resources and goods that *would* ensure their survival. As a result, Winstanley argued that 'if the waste land of England was manured . . . it would become in a few years the richest, strongest and [most] flourishing land in the World'.[67] Once the environment was unlocked and all were free to labour for themselves rather than for hire, there would be enough for all.

Second, Winstanley held that this alternative system would secure universal personal independence and self-sufficiency. By labouring on their own resources, under their own direction, all could support themselves without recourse to the land owners, clergy and lawyers. Encouraging the English people to follow the example of the Diggers, he asked, 'will you be slaves and beggars still, when you may be freemen? Will you live in straits and die in poverty, when you may live comfortably?'[68] He argued that instead of the 'oppression' they currently suffered under 'you may be free' because 'true freedom lies where a man receives his nourishment and preservation and this is in the use of the earth'.[69] Creating what they needed to support themselves, needing neither permission nor help from any but God, would enable all to live freely and equally, in contrast to the unequal government that resulted from private ownership of the earth.

The alternative society Winstanley imagined was founded on the shared ownership of the environment, maintained through and designed to enable the ability of all to labour upon it. This would create enough resources for all and enable them to live independently. Natural resources had been the means through which the powerful controlled and governed the majority – through labouring upon the earth, the poor and oppressed could reclaim these resources and use them to create their own alternative, equal society.

Action joined with thought: Winstanley and praxis

There is a final point to be made regarding Winstanley's engagement with the concept of labour and its links to governing the earth. The opening section introduced Winstanley's 'vision' to begin digging on St George's Hill, the development of the Digger movement and its eventual disbandment in the face of (frequently violent) opposition and then explored the thought and writing that was the impetus for and tied to this action. This section was structured to reflect the centrality of physical action. When Winstanley experienced his vision, his response was physical, moving to St George's Hill and beginning labouring upon the land, not just to write and publish the message he received. His writing from 1649 onwards sought primarily to document, justify, publicise and defend these actions ('therefore we justify our act of digging upon that hill to make the earth a common treasurie').[70] This emphasis on physical action over writing is epitomised in *A Watch-Word to the City of London and the Army* when he notes that 'yet my mind was not at rest because nothing was acted and thoughts run in me that words and writings were all nothing, and must die, for action is the life of all and if thou dost not act, thou dost nothing'.[71]

Winstanley held that words without action were nothing for two reasons. The first relates to his experience of bankruptcy. Winstanley went bankrupt in 1643 whilst working as a cloth-merchant in London, and the experience left him with a hatred of business, especially those who made promises and agreements that they would not keep.[72] Even when trying to seek support for the Digger movement from the City of London, his bitterness at those 'cheating sons in the thieving art of buying and selling' comes through.[73] Second, Winstanley argued that the Parliamentarian victory in the British Civil Wars had been achieved through the help of the poorest, who had risked and given their lives as well as paid 'free-quarter' in the support for the Army. The people, he argued, had joined the war in order to defeat 'kingly power' and the Norman yoke and been promised their freedom in return, but landlords and private property remained.[74] Winstanley supported swearing allegiance during the Engagement Crisis, but this only fuelled his resentment that the war had merely replaced the figurehead of power rather than tearing down the system of government. Words had been plentiful, but they had not been followed with action.

This link between action and thought reflects the contemporary concept of praxis, which is the link between the two that results in informed action. This aspect of Winstanley's work has been recently explored. Thomas H. Corns, for example, builds on Winstanley's statement that 'I have writ, I have acted, I have peace' to note that only through writing and acting together could Winstanley feel that he had fulfilled his vision.[75] Corns points out that 'for Winstanley the dynamic is a familiar one, the assertion of that praxis in which the theory in his writing is enacted in political action, with the comfort that comes from knowing as much as has been done as possible'.[76] Christopher Rowland also notes that 'the centrality of experience' in Winstanley's work is 'striking',[77] though he is less confident of using the term *praxis*.[78] Yet Rowland does state that 'it may be appropriate for a modern commentator to point out the references to the dire social and economic context in Winstanley's writings, which suggests he saw a link between "life" and "writing"'.[79]

This chapter supports the 'appropriateness' of this approach in two ways, both of which focus on the 'earthly' facets of this concept. The first is through a focus on the temporal aspects of praxis. The links between Winstanley's religious thought and the praxical approach have been explored by other commentators, but the focus here is on the application to the other elements of his work. This analysis therefore complements rather than rejects the work of Rowland. Second, the concept of praxis is connected to his repeated emphasis on the importance of labour upon natural resources. As has been shown, Winstanley saw labour as central to changing 'the dire social and economic circumstances' of his time.[80] Winstanley did not only argue that those who were shut out of resources had been wronged, he argued that they should take action to change their circumstances and demonstrated this action himself. The understanding of Winstanley's suggested solution as a conscious redirection of labour rather than a strike underlines this conclusion – this is about choosing to practice certain kinds of action (as a result of theoretical arguments) rather than a withdrawal from action. In his discussion of Winstanley's connection between theory and practice, Corns rightly highlights the importance of his statement that 'I have writ, I have acted, I have peace'; however, that 'act' is specifically that of labouring upon natural resources.[81] Winstanley's praxis is an 'earthly' one, for it concerns the worldly aspect of his work as well as the spiritual and is based upon a directed, deliberate engagement with the earth and other natural resources.

This aspect of Winstanley's politics highlights the role of the body as a site and agent of resistance and in doing so emphasises the embodied physicality of governing and political theorising.[82] Winstanley held that labour was the source of political resistance, the means to reshape society. And given the context, this is physical labour: ploughing, digging, sowing, reaping and chopping. Winstanley identified the labour of the body as the means by which the poorest could transform society, in part because it was the tool most easily available to them, especially when the law courts and parliamentary system seemed, to Winstanley, to be closed under the 'Norman yoke'. Winstanley implored the people of England to use their bodies to transform their environment: 'to stand up for your freedom in the land, by acting with plow and spade upon the commons', asking 'will you not rise up and act . . . but come take plow and spade, build and plant and make the wasteland fruitful'.[83] Therefore, in Winstanley's writing we see the relationship between government, the earth and all its resources, and the body, with the dependence of the human body upon the natural world enabling a minority to govern and maintain power through controlling access to those resources. Yet the body is a source and means of labour, which can in turn transform these resources and so reshape both the earth and its government. The body's essential connection to the environment creates a vulnerability that can be exploited, but for Winstanley this connection also created the necessity for political change and provided the means by which to achieve it.

We think and act politically whilst being present in the environment, arguing about the ownership of resources whilst breathing them in and existing in that space around us. Looking to Winstanley's work makes this clear, as he brings the body's relationship with the earth to the fore. Our government is embodied and dependent on the environment – we govern the earth whilst living in and on it, unable to separate ourselves from the resources we seek to control.

Conclusion

This chapter has shown that the concept of labour is central to Winstanley's understanding of how the natural resources of England *were* governed, how they *could* be governed and the transition between the two. After introducing Winstanley's thought and actions, the second section examined Winstanley's argument that England was governed through the environment with private ownership of land and other resources, backed by the power and legitimacy of the law and the Church, held to be central to the ability of an elite to control England. Only through their labour could 'the poor and oppressed peoples of England' hope to challenge this control and take back what was rightfully theirs. Through refusing to work on privately owned resources and instead deliberately choosing to redirect their labour onto the common lands, Winstanley argued that those who did not own or govern resources could become independent and self-sufficient whilst undercutting the existing power structure. Understanding Winstanley's plan as a re-direction of labour shows why this is the means through which he thought control of political power could be seized and an alternative form of government, one not based on 'kingly power' and the Norman yoke but on equal control of the earth, could be established. The final section argued for the centrality of labour within Winstanley's political activity, arguing that the link between his action and his writings, his digging the earth and the writings that result, reflect the concept of praxis. Though the praxical elements of Winstanley's work have recently come under examination, the explicit link to labouring upon the environment has not been explored, the final section addresses this and considers the embodied government that results.

The introduction outlined some criticisms of an environmental reading of Winstanley. By specifically focusing on the role of labour in the preceding sections, this chapter has endeavoured not to fall into the same trap. The resulting understanding confirms Hessayon's point that Winstanley cannot fit comfortably within a Green narrative due to his emphasis on the large-scale use and development of resources. Ken Hiltner's explanation of why the Diggers faced opposition from the local poor is also applicable here – Hiltner notes that 'it would hardly be surprising if, in spite of his many protestations regarding his concern for the earth, Winstanley was thought by these people to be just another projector who wanted to radically alter their common for a new use'.[84] Winstanley believed that the earth had been given to all by God specifically so that they could extract the materials they required and transform these resources to better fit their needs – a belief few contemporary environmentalists share, making him another past thinker who wished to 'radically alter' and exploit the earth but linking him to the larger history of English radicalism.[85]

However, this could be a productive tension. First, Winstanley's praxis and his focus on the body, his physical embodiment of his writings draws attention to the need to match Green theory with action and to the fact that that we theorise about our relationship with our environment and with other people through it whilst embedded within the natural world. When seeking to bring about a change in the current distribution of the earth and all its resources, the body may be the best tool available. This recasts the relationship between labour and the environment as a more mutual one (or certainly one with the *potential* to be mutual) rather than inherently antagonistic.[86] As a result, Winstanley's work and action support the arguments of environmental historians of work such as Thomas Andrews, who suggest the concept of the 'workscape' to highlight the intersection of the landscape and labour.[87]

This ties to the second point, which is that Winstanley shows the importance of labour upon the environment both as a means of resistance and as a source of alternative. His work draws attention to the benefits of labouring upon resources in order to ensure the independence and survival of those who are dispossessed and disadvantaged. An understanding of environmental relationships that is centred on labour and most importantly equitable, shared labour may seem contradictory, but it presents a fascinating tension, a way of challenging and creating alternative governments that are accessible to all. Thus, Winstanley's emphasis on the conditions under which this development takes place is crucial for thinking about environmental justice. Recognising and acknowledging the critical role of labour in the nexus of land and power will therefore create a different kind of environmentalism, one which is not torn by the tensions identified by commentators such as Richard White.[88] As a result, understanding the link between labour and control of the earth presented here is not just important for understanding the thought and action of Gerrard Winstanley and of the conceptions of resource management control and resistance in the early modern period. It is important for any understanding of how the earth is governed, for what purpose, and by whom.

Notes

1 Thanks are due to the participants at the Ruling Climate conference, and this chapter has benefitted greatly from the comments and advice of the editors and the anonymous reviewer.
2 Mark Kishlansky, quoted in T. Corns, A. Hughes and D. Loewenstein, 'Introduction', in *The Complete Works of Gerrard Winstanley*, ed. T. Corns, A. Hughes and D. Loewenstein, 2 vols. (Oxford: Oxford University Press, 2009), i, 1.

3 G. Kennedy, *Diggers, Levellers and Agrarian Capitalism* (Plymouth: Lexington Books, 2008); J. Gurney, *Gerrard Winstanley: The Digger's Life and Legacy* (London: Pluto Press, 2013); Corns, Hughes and Lowenstein, 'Introduction', 46–47.
4 See T. Benn, *Gerrard Winstanley: A Common Treasury* (London: Verso, 2011), G. Monbiot, 'Still Digging', June 2000. http://www.monbiot.com/archives/2000/06/01/still-digging/. Accessed 13 June 2016; R. Spowers, *Rising Tides* (Edinburgh: Canongate, 2002), 109 for examples. The trend is also highlighted in J.D. Alsop and J.C. Davis, 'Winstanley, Gerrard, (*bap.* 1609, *d.* 1676)', *Oxford Dictionary of National Biography*, 60 vols. (Oxford: Oxford University Press, 2004), lix, 762–770.
5 A. Hessayon, 'Restoring the Garden of Eden in England's Green and Pleasant Land: The Diggers and the Fruits of the Earth', *Journal for the Study of Radicalism* 2, no. 2 (2008): 17.
6 Ibid.
7 G. Winstanley, 'The New Law of Righteousness, 1649', in *The Complete Works of Gerrard Winstanley*, ed. Corns, Hughes and Loewenstein, i, 472–568; Winstanley, 'A Declaration to the Powers of England (The True Levellers Standard Advanced), 1649', in *The Complete Works of Gerrard Winstanley*, ed. Corns, Hughes and Loewenstein, ii, 1–20.
8 Gurney, *Gerrard Winstanley*, 47; Winstanley, 'A Declaration to the Powers of England', 13; Winstanley, 'A Watch-Word to the City of London and the Army', 1649, in *The Complete Works of Gerrard Winstanley*, ed. Corns, Hughes and Loewenstein, ii, 79–106.
9 Corns, Hughes and Lowenstein, 'Introduction', 28.
10 Ibid., 30.
11 Gurney, *Brave Community*, 143–167; Gurney, *Gerrard Winstanley*, 69.
12 Gurney, *Brave Community*, 166.
13 Winstanley, 'A Watch-Word to the City of London and the Army', 90.
14 Gurney, *Brave Community*, 195.
15 The opening direction to Oliver Cromwell in dated by Winstanley to November 1651 (Winstanley, 'The Law of Freedom in a Platform', in *The Complete Works of Gerrard* Winstanley, ed. Corns, Hughes, and Lowenstein, ii, 278–404), and Winstanley most likely worked on the text during the preceding years. However, 'the tract does not seem to have been published until February 1652' (Gurney, *Brave Community*, 386, note 52).
16 Gurney, *Gerrard Winstanley*, 86.
17 Corns, Hughes and Lowenstein, 'Introduction', 18.
18 Winstanley, 'An Humble Request to the Ministers of both Universities and to All Lawyers in Every Inns-a-Court', in *The Complete Works of Gerrard Winstanley*, ed. Corns, Hughes and Lowenstein, ii, 255–277; Winstanley, 'A Watch-Word to the City of London and the Army', 85. See also P. Harrison, 'Fill the Earth and Subdue It: Biblical Warrants for Colonization in Seventeenth Century England', *Journal of Religious History* 29, no. 1 (2005), 12.
19 Winstanley, 'The New Law of Righteousness', i, 520; Winstanley, 'An Appeal to the House of Commons', 73.
20 C. Hill, *The World Turned Upside Down: Radical Ideas during the English Revolution* (London: Penguin, 1972), 145; L. Brace, *The Politics of Property* (Edinburgh: Edinburgh University Press, 2004), 20.
21 Winstanley, 'A Declaration to the Powers of England', 5–8.
22 The word yoke is spelt as 'yoak' by Winstanley – the modern spelling is used here for convenience and clarity.
23 J. Gurney, 'Gerrard Winstanley and the Concept of Place', *Prose Studies* 36, no. 1 (2014): 9. For examples, see Winstanley, 'A Declaration to the Powers of England', 12–13; Winstanley, 'An Appeal to the House of Commons', ii, 71–72; Winstanley, 'A Watch-Word to the City of London and the Army', 92–93.
24 Winstanley, 'An Appeal to All Englishmen', in *The Complete Works of Gerrard Winstanley*, ed. Corns, Hughes and Lowenstein, ii, 243–254.
25 Winstanley, 'The Law of Freedom in a Platform', ii, 302.
26 Winstanley, 'A Declaration to the Powers of England', 11; Winstanley, 'The Law of Freedom in a Platform', 296.

27 Winstanley, 'A Declaration from the Poor Oppressed People of England', in *The Complete Works of Gerrard Winstanley*, ed. Corns, Hughes, and Lowenstein, ii, 31–42.
28 Ibid.
29 Winstanley, 'The Law of Freedom in a Platform', 359.
30 Ibid., 373.
31 Ibid., 323–329.
32 Ibid., 328, 378.
33 D. Webb, 'The Bitter Product of Defeat? Reflections on Winstanley's Law of Freedom', *Political Studies* 52 (2004): 199–215.
34 Corns, Hughes and Lowenstein, 'Introduction', 45.
35 This was also supported by the established church through their hierarchical approach to religion, which held priests had access to religious truths others did not, and the imposition of tithes (see Corns, Hughes and Lowenstein, 'Introduction', 51–59 and Winstanley, 'The Law of Freedom in a Platform', 282 for but one example).
36 Winstanley, 'Fire in the Bush', in *The Complete Works of Gerrard Winstanley*, ed. Corns, Hughes, and Lowenstein, ii, 171–234.
37 J. Locke, *Second Treatise of Government*, ed. C.B. Macpherson (Indianapolis: Hackett, 1980), 26.
38 R. Williams, *Problems in Materialism and Culture* (London: Verso, 1980), 76, 79–80.
39 P. Harrison, 'Fill the Earth and Subdue It', 12.
40 See for further A. McRae, 'Husbandry Manuals and the Language of Agrarian Improvement', in *Culture and Cultivation in Early Modern England: Writing and the Land*, ed. M. Leslie and T. Raylor (Leicester: Leicester University Press, 1992), 45; P. Warde, 'The Idea of Improvement, c.1520–1700', in *Custom, Improvement and the Landscape in Early Modern Britain*, ed. R. Hoyle (Farnham: Ashgate, 2011), 140.
41 Corns, Hughes and Lowenstein, 'Introduction', 41, 46; K. Hiltner, *What Else Is Pastoral?* (Ithaca: Cornell University Press, 2011), 128; P. Slack, *The Invention of Improvement* (Oxford: Oxford University Press, 2015), 94. There is also a link here to the work of William Covell, who suggested the development of a colony at Enfield in 1659 that reflected the Digger aims – see B. Sharp, 'Rural Discontents and the English Revolution', in *Town and Countryside in the English Revolution*, ed. H.C. Richardson (Manchester: Manchester University Press, 1992), 268.
42 D.W. Petegorsky, *Left-Wing Democracy in the English Civil War* (London: Gollancz, 1940); Corns, Hughes, and Lowenstein, 'Introduction', 27–28; Gurney, *Brave Community*; Gurney, *Gerrard Winstanley*.
43 Winstanley, 'The Law of Freedom in a Platform', 289.
44 Winstanley, 'A Watch-Word to the City of London and the Army', 97.
45 N. Blomley, 'Making Private Property: Enclosure, Common Right and the Work of Hedges', *Rural History* 18, no. 1 (2007), 1–21.
46 Winstanley, 'An Humble Request to the Ministers of both Universities and to All Lawyers in Every Inns-a-Court', 258.
47 Winstanley, 'An Appeal to All Englishmen', 244.
48 Alsop and Davis, 'Winstanley, Gerrard', 766.
49 Winstanley, 'A Declaration to the Powers of England', 11.
50 Winstanley, 'The New Law of Righteousness', 516–517.
51 Winstanley, 'A Declaration to the Powers of England', 16.
52 G. Kennedy, *Diggers, Levellers and Agrarian Capitalism* (Plymouth: Lexington Books, 2008), 129.
53 Hill, *The World Turned Upside Down*, 131; Gurney, *Gerrard Winstanley*, 45.
54 Winstanley, 'A Declaration from the Poor Oppressed People of England', 35.
55 Gurney, *Gerrard Winstanley*, 46; Gurney, *Brave Community*, 102.
56 Winstanley, 'A Watch-Word to the City of London and the Army', 88. Such far-reaching and expansive interpretations of the right of common have been present and key in the radical movements since the seventeenth century. B. McDonagh and C. Griffin, 'Occupy! Historical Geographies of Property, Protest and the Commons, 1500–1850', *Journal of*

Historical Geography 53 (2016): 5. See A. Howkins, 'From Diggers to Dongas: The Land in English Radicalism, 1649–2000', *History Workshop Journal* 54 (2002): 1–23.

57 C. Hill, 'Introduction', in *Winstanley: The Law of Freedom and Other Writings*, ed. C. Hill (Cambridge: Cambridge University Press, 1983), 40.

58 Winstanley, 'A New-Years Gift for the Parliament and Army', 133.

59 Winstanley, 'A Watch-Word to the City of London and the Army', 95.

60 Elaine Hobby has however questioned the extent of the role of women within both the Digger community and Winstanley's writings, and her resulting argument questions whether women's equality was acknowledged. See E. Hobby, 'Winstanley, Women and the Family', *Prose Studies* 22, no. 2 (1999): 61–72. For an alternative view, see Kennedy, *Diggers, Levellers and Agrarian Capitalism*, 198.

61 Winstanley, 'The Law of Freedom in a Platform', 371.

62 Ibid., 291.

63 Webb, "The Bitter Product of Defeat?'.

64 Winstanley, 'Fire in the Bush', 223.

65 Winstanley, 'The Law of Freedom in a Platform', 285.

66 D. Russell, 'Locke on Land and Labour', *Philosophical Studies* 117, nos. 1–2 (2004): 309.

67 Winstanley, 'An Appeal to All Englishmen', 244.

68 Ibid., 243.

69 Winstanley, 'The Law of Freedom in a Platform', 296, 247, 295.

70 Winstanley, 'A Watch-Word to the City of London and the Army', 85; see also Winstanley, 'A Vindication of Those, Whose Endeavours Is Only to Make the Earth a Common Treasury, Called Diggers', in *The Complete Works of Gerrard Winstanley*, ed. Corns, Hughes, and Lowenstein, ii, 235–242.

71 Winstanley, 'A Watch-Word to the City of London and the Army', 80.

72 Gurney, *Gerrard Winstanley*, 17; J.D. Alsop, 'Ethics in the Marketplace: Gerrard Winstanley's London Bankruptcy, 1643', *Journal of British Studies* 28, no. 2 (1989), 97–119.

73 Winstanley, 'A Watch-Word to the City of London and the Army', 80.

74 For example, Winstanley, 'The Law of Freedom in a Platform', 281.

75 Winstanley, 'A New-Years Gift for the Parliament and Army', 149.

76 T. Corns, 'I Have Writ, I Have Acted, I Have Peace: The Personal and the Political in the writings of Winstanley and Some of His Contemporaries', *Prose Studies* 36, no. 1 (2014): 46. See also Corns, 'Christopher Hill on Milton, Bunyan and Winstanley', *Prose Studies* 36, no. 3 (2014): 215, in which he notes that 'Winstanley's theology advances a powerful dialectic between theory and practice'.

77 C. Rowland, 'Gerrard Winstanley: Man for All Seasons', *Prose Studies* 36, no. 1 (2014): 80.

78 Ibid., 87.

79 Ibid.

80 Ibid.

81 Winstanley, 'A New-Years Gift for the Parliament and Army', 149.

82 For embodiment, see T. Benton, *Natural Relations: Ecology, Animal Rights and Social Justice* (London: Verso, 1993), 103.

83 Winstanley, 'An Appeal to All Englishmen', 244, 248.

84 Hiltner, *What Else Is Pastoral?*, 129.

85 Howkins, 'From Diggers to Dongers'.

86 Reflecting the argument of Richard White, 'Are You an Environmentalist or Do You Work for a Living? Work and Nature', in *Uncommon Ground: Rethinking the Human Place in Nature*, ed. William Cronon (New York: W.W. Norton & Company, 1997), 171–185.

87 Thomas G. Andrews, 'Work, Nature, and History: A Single Question, That Once Moved Like Light', in *The Oxford Handbook of Environmental History*, ed. Andrew C. Isenberg (Oxford: Oxford University Press, 2014).

88 White, 'Are You and Environmentalist or Do You Work for a Living?'.

6 Winter and discontent in early modern England

William M. Cavert

John Evelyn's diary for the winter of 1683/4 records remarkable scenes in London. A spell of extreme cold had filled the city's air with stagnant coal smoke that rendered breathing nearly impossible, the river Thames was frozen 'by extreme hard weather,' and streets of booths and shops were sprung up on the ice. As England's foremost enthusiast for and expert on gardening, Evelyn lamented that so many of the trees and plants in his extensive gardens were 'utterly destroyed.' As a former office holder and rather stern critic of public morality, he was unimpressed by the 'bull baiting, horse and coach races, puppet plays and interludes, cooks and tippling, and other lewd places' that he found so quickly built on London's frozen river. As a believer in providence, he read the suffering of the poor as a commentary on the nation's sins: 'it seemed to be a bacchanalia, or triumph, or carnival on the water, whilst it was a severe judgment on the land.'[1]

Evelyn's concern for the plight of the poor and of his fruit trees was very different from the exaltation or exasperation he had displayed in response to other events during England's century of revolution. He praised God in gratitude for Charles II's Restoration of 1660, but by 1666 he had turned against 'the vanity and vices of the court, which makes it a contemptible thing.'[2] His widely quoted descriptions of the Great Fire of 1666 are full of denunciations of moral and political failures, and he found the Popish Plot and associated crises of 1679 to be 'public confusions in church and kingdom, never to be sufficiently deplored.'[3] The winter of 1683/4, by contrast, did not seem to call for such commentary. Instead of interpreting the hard freeze as a political crisis, Evelyn's response, like that of many of his contemporaries, was ambivalent. There was real suffering yet also undeniable fun, a kind of battle between the principles of carnival and Lent played out, not on a canvas by Breughel but on the streets and in the houses of London.

Historians have recently introduced another way of reading extreme weather events during the early modern period. During the Little Ice Age, which endured roughly from 1300–1800 (but with its precise timing and extent, and the extent of fluctuations within this period, disputed), western Europe experienced colder average temperatures than during the previous late-medieval period. This led to reduced growing seasons as well as several particularly cold periods. Explaining the consequences, both material and cultural, of these processes is a hugely challenging project, but one that has recently been taken up from a multitude of perspectives.[4] One prominent strand of this literature has been to use climate to explain what would otherwise remain a remarkable coincidence, the nearly contemporaneous outbreak of civil wars and rebellions across the Eurasian world in the 1640s – the General Crisis of the Seventeenth Century. Winters such as 1683/4, however, raise an important limitation of this approach, since understanding the cultural history of the early modern climate requires not only knowledge of climatic processes, growing seasons, and crop yields but also close attention to how weather trends and events were explained and constructed and influenced behaviors and choices. Such attention shows that during England's century of revolution, the problems raised by extreme winters were not thought to threaten regimes but rather, to the extent that they mattered to the regime at all, to allow them to demonstrate the benevolence of their rule.

Climate and the General Crisis of the 17th century

The 'General Crisis' thesis arose during the 1950s and sparked fierce debate between Marxist and anti-Marxist scholars of European history, led in Britain by Eric Hobsbawm and Hugh Trevor-Roper.[5] Since then, however, British historians have found it possible to debate the origins, course, and significance of what is variously called the English Civil War, English Revolution, Puritan Revolution, or War of the Three Kingdoms with almost no reference to this debate at all.[6] Outside of British historiography, the General Crisis thesis fared only slightly better, lying in 'the historiographical dustbin' for decades before becoming, since 2008, 'the crisis that will not go away.'[7] The work most responsible for that resurgence has been from Geoffrey Parker, whose 2008 article in the *American Historical Review* and subsequent book, *Global Crisis: War, Climate Change and Catastrophe in the Seventeenth Century*, argued explicitly for a causal connection between the Little Ice Age and the General Crisis.[8] Others have recently applied Parker's approach to specific cases, with varying degrees of success, but *Global Crisis* remains by far the most ambitious attempt to connect the early modern climate to a global (or at least pan-Eurasian) political narrative spanning a century.[9]

Parker argues that climate influenced politics in two ways, which the narrative often intertwines but are nevertheless conceptually separate. The first is a materialist argument, in which colder weather led to generally short growing seasons as well as to years of particular stress (like late frosts, early winters, summer droughts) that in turn led to dearths, famines, and diseases. Such failures of the agrarian economy reduced the wealth that could safely be extracted by rulers. Climate change here leads directly to economic and therefore fiscal crises and therefore also to political crises. Japan alone successfully weathered these challenges through a series of enlightened policies.[10] This aspect of the book has attracted the most attention from reviewers, as economic historians have challenged this materialist story from every angle. Paul Warde, for example, argues that Parker's narrative is essentially 'good old-fashioned Malthusianism' but lacks data to establish 'clear-cut demonstrations of cause and effect.'[11] Kenneth

Pomeranz challenges key aspects of his treatment of Asian developments.[12] Jan de Vries goes so far as to assert that 'economic history' itself is entirely absent from Parker's account, asserting that economic performance is reduced to a mere epiphenomenon of weather.[13]

Economic historians, therefore, are not convinced whether or how early modern climate change impacted economic performance. But even if it *could* be conclusively demonstrated that long-term processes or shorter-term weather events were in fact harmful to early modern economies, it does not necessarily follow that such stresses would entail problems for rulers and states. Whether or not they did so is an empirical question, and in the case of the English Civil War of the 1640s, there are good reasons to doubt this connection. Parker's narrative draws heavily on what early modern British historians call the 'revisionist' approach to early-Stuart politics. While there is diversity within this school, all of its major protagonists' major publications, from the 1970s through the early 1990s, shared characteristics that influence Parker's account. They downplay ideological conflict based on differing interpretations of the English constitution, political rights, and the rule of law. Instability arose, instead, from structural problems besetting the English state, particularly the rising real cost of warfare coupled with a creaky and unprofessional administrative apparatus that made it difficult for England's rulers to extract wealth from their people.[14] The other crucial structural problem was the composite nature of the English-Scottish-Irish crown. This is 'the British problem,' in which England's kings were weakened rather than strengthened by the fact that they also ruled the subsidiary crown of Ireland and the entirely independent and separate crown of Scotland. Being a 'composite monarchy' was particularly difficult because the differing legacies of the Reformation in each country meant that James I and Charles I could not pursue uniform religious policies across their kingdoms, but nor could they allow subjects freedom of worship outside of a national church. James I, as a skillful political operator, managed to prevent these tensions from coming to a head. But his son and successor Charles was far less crafty, far too inflexible, and far too naive politically to avoid falling into the traps laid by these structural weaknesses. England's civil war, and the broader British wars of which it was a part, were therefore caused not by constitutional conflict but by structural tensions, religious prejudice, and the contingent fact that Charles I was temperamentally 'unfit to be king.'[15]

During the past thirty years, however, the insufficiencies of this formulation have been stressed by a range of studies clumsily labeled as 'post-Revisionist'. They argue, among other things, that political conflict, including constitutional disagreement, was widespread during the early seventeenth century, and that this – along with more contingent opposition to specific political choices by both James and Charles – meant that what the Revisionists took to be structural features were in fact political. Even the nature and practice of political debate was the subject of political debate, as politicians alternately used and denounced the public sphere of the pulpit, print media, and drama. Religion, they stress, was not merely unthinking and inexplicable prejudice. Rather it was deeply and intrinsically political, raising and complicating secular concerns related to the constitution, rule of law, and nature and limits of kingship and obedience.[16] None of this is part of Parker's story, which, historiographically, is a late addition to the revisionist literature. The post-revisionist critique of this position, however, is a problem for the climate-centered thesis he advances because it is increasingly difficult to claim that war resulted primarily from structural problems. If the English civil war was the result of explicit and conscious political divisions, of clashing ideas, beliefs, and ambitions, then it is unclear why a series of bad harvests or high prices would help to bring about such an ideological war. If bad weather or longer-term climatic changes *did* undermine the strength of the Stuart regime and allow for civil war, it is not currently clear how this worked.

An additional critique of this materialist argument emerges from literature on riots, dearth, and popular politics. As part of his broader project to reinsert the poor into the political nation, E.P. Thompson argued in 1971 that hunger is not a sufficient explanation for a food riot. Of course people naturally need food, he conceded, but we should not presume to know how bodily appetites will be manifested in behaviors, since such actions arise not merely from sensory impulses but also from the more complex urges and constraints of culture and politics. Rioters, he argued, were not merely hungry, and were certainly not a mindless horde, but rather were seeking to require their governors to govern properly, in accordance with the needs of poor consumers, and to curb the unjust profiteering of middlemen. The broader point of his 'moral economy of the crowd' is that riots were about perceptions of injustice and bad governance as much as they were about hunger.[17]

This is a key point, made repeatedly over the past forty years by several social historians, many of whose work, like Thompson's, is otherwise keen to stress the ambitions, agency, and activity of the poor.[18] The objectives of riots tended to be limited, focusing on specific and fairly uncontroversial claims such as the right to subsistence or to enjoy customary (and therefore legally binding) access to resources. They were, as Keith Wrightson wrote in 1982, 'not the desperate and bloody furies of a demoralized mob. Whatever their menacing postures, rioters were concerned with the rectification of specific grievances, and infused with a sense of justice.'[19] Rioters were politically conscious actors striving for specific ends rather than pitchfork-wielding peasants wantonly murdering lords and burning manor houses. As such, they were at least as likely to invoke, appropriate, or call on state power as they were to undermine or overturn it.[20] Whereas Parker cites Francis Bacon's dictum that 'rebellions of the belly are the worst' as a wise insight into early modern politics, scholars of popular politics have shown this to be rather an elite fantasy, projecting a normative model of political quiescence onto poor people who were, in fact, capable of rational and limited political actions under a variety of circumstances, not merely when they were driven to desperation by hunger. Empty bellies, in short, did not lead to rebellion in early modern England.

Recent studies of rebellions, moreover, similarly stress the sophistication of their political motivations and maneuvers. The risings and 'commotions' of 1549 and the Northern Rising of 1569, for example, have been analyzed according to the precise political goals of the participants.[21] In some cases these goals are economic, in some cases ideological or religious, and in some cases these categories were entangled. But Parker's claim that rebellions were primarily motivated by the desperate and immediate fear of starvation finds little support in this literature. In particular, such a connection is absent from the leading study of the rebellion that is at the heart of Parker's story. The Scottish 'Prayer Book Rebellion,' which is the case study chosen for the 2008 *American Historical Review* article, is presented as deriving from a climate-induced harvest crisis.[22] Charles I's plan to introduce a new liturgy in Presbyterian Scotland was doomed because it was introduced in the summer of 1637, 'a dangerous moment to innovate because Scotland suffered more extreme weather than England.'[23] Yet recent research has found no such evidence at all connecting weather or harvests and these riots. They were urban rather than rural, they attacked signs of ecclesiastical authority rather than grain stores, they wanted to alter religious policies rather than lower prices, and they were carried out by paid retainers of powerful lords rather than starving peasants. In short, they were not 'motivated by economic suffering and hardship.'[24] The Edinburgh rioters of 1637 therefore may have been the people in Scotland *least* impacted by bad weather or poor harvests.

The structural model, by which bad weather led to political instability in Britain, is therefore seriously flawed. There is also, however, another strand to the argument. This is what Parker calls the 'peccatogenic outlook,' or more simply the 'search for scapegoats.'[25] Because God took a keen interest in human affairs and human morality, the sins of individuals, rulers, and entire polities were subject to divine or heavenly punishment. Extreme weather events, earthquakes, comets, and eclipses were believed to signal divine displeasure. One advantage of this approach for a global argument is that it is at once culturally specific, since European Catholics or Protestants and Chinese Buddhists or Confucians might read the heavens very differently, and yet also can be applied across Eurasia and across the period so as to connect almost any weather event or climatic process with a subsequent period of instability. Another benefit of this part of the narrative is that it – unlike the more 'Malthusian' aspects of the argument – is not dependent on reconstructing the actual extent of the early modern Little Ice Age, a process that scientists continue to study and which a few historians have even questioned altogether.[26] Even if the idea of a Little Ice Age were a historian's meaningless invention based on flawed research (which it is not), Parker's peccatogenic outlook would still explain how bad weather could lead to political instability. A drought, frost, or famine could still suggest to early modern believers some form of divine judgment, even if climatologists consider such an event a mere short-term blip with no long-term significance. In this model it was the ideas of early modern people, not the objective conditions of climate change, that mattered most.

A fundamental part of this argument is that divine judgment was believed to have been directed towards regimes and rulers, those whose shortcomings had incurred God's wrath or lost them the mandate of heaven. Similar work linking environmental with cultural change includes the work of Wolfgang Behringer, who has attributed the central European witch hunt, and many other cultural developments besides, to shockingly cold new weather in late-sixteenth century southern Germany.[27] Like Behringer, Parker finds that weather anomalies were clearly evident to contemporaries and prompted them to seek explanations, a process that led them to question the legitimacy of their rulers and, ultimately, to revolt against them.[28]

While reviews by economic historians have focused on the structural aspect of Parker's argument, this cultural strand has received far less attention. In a basic sense it is clearly true that, in the British case as elsewhere, early modern people believed in providence and used it to explain events. Alex Walsham's work, in particular, demonstrates how widely providentialism was shared in early modern England; the question for contemporaries was not whether but how god's power and wisdom were manifested in particular events.[29] The ubiquity of providentialism in early modern Europe thus led not to consensus but to another way to articulate conflict, as rival confessions and political groups offered sharply opposing interpretations of storms, floods, and other omens.[30] To what extent disasters were judgments at all was also part of this disagreement; depending on the polemical, political, and institutional context, a natural event like a flood or a crop failure could variously be attributed to divine wrath, poor preparation by the government or community, natural variation, or some complex combination of such factors.[31] It would therefore be surprising to find people in early modern Britain unequivocally using weather events to claim that Charles I's regime was no longer authorized by God; studies from Britain and elsewhere in Europe suggest that opposing readings were at least as likely, and recalling the treasonous nature of such claims further helps to explain the absence of documents supporting the suggestion that Charles I's subjects blamed him for difficult weather.

The Great Frost of 1684

The Little Ice Age, as a period in the history of European climate, is primarily defined by average temperature levels over the long term, within which were fluctuating periods of colder or more moderate trends.[32] Among its most characteristic manifestations, however, and among those most perceptible on human time scales to people living in the northern hemisphere, were its many hard frosts and deep freezes.[33] While an unusually cold winter may well be followed by a pleasant summer and a bountiful harvest and therefore may not lead to any of the structural problems stressed by Parker, they were exactly the kind of anomalous weather that the 'peccatogenic outlook,' according to his formulation, should attribute to bad governance. If contemporaries accepted that such 'extreme weather events' demonstrated – or even hinted – that the regime had lost its legitimacy, then we would expect to find such beliefs reflected in the sources. Examining the coldest recorded winter of the period, however, shows something else.

The coldest winter of the Little Ice Age in England, indeed the coldest winter in that country's entire history since the beginning of continuous record keeping in 1659, occurred in 1683/4. From December through February its average, according to a series compiled by Gordon Manley, was -1.2°C.[34] This is .8°C colder than the second-coldest known winter (1739–1740), and over 6°C colder than the first fifteen English winters of the twenty-first century.[35] The winter of 1683/4 was, in fact, famously cold, as the Thames frost fairs were repeatedly illustrated and are today inevitably reproduced in visual and digital media during any cold or snowy period. It was, for many, 'the Great Frost,' the worst winter in anyone's memory. According to printed broadsheets, pamphlets, and sermons, it was a 'wonder,' a unique – or at least extremely unusual – event calling out for explanation and explication.[36]

But unlike the bad years of the late 1630s, the early 1640s, and the late 1640s, 1683/4 was emphatically not a moment of political revolution in England. To the contrary, after a tumultuous period from 1679–1681 when the country seemed very much on the brink of civil war, by 1683 the political momentum had shifted unmistakably in favor of the Stuart crown. A plot amongst leading oppositionist Whigs was detected, or perhaps fabricated, and from the summer of 1683 the regime steadily advanced against its enemies. The Whig leader Shaftesbury had already gone into exile, and he was soon followed by his protégé John Locke. Many of those implicated in the plot, including prominent figures like the Earl of Essex, Lord Russell (eldest son of the Earl of Bedford), and the prominent intellectual Algernon Sidney, were judicially executed or died in prison. By winter, not only were actual conspirators crushed, but those who had defended them or questioned the case against them were similarly suppressed. Borough after borough, including the City of London itself, lost its charter and thus its ability to elect a Member of Parliament unpalatable to the crown. The king's illegitimate son and the sometime favorite of disaffected Protestants, the Duke of Monmouth, was received at court. He was forgiven for suspicions of involvement in the Plot but thereafter kept at arm's length, neither powerful enough to protect the king's enemies nor desperate enough to pursue rebellion. Oxford University publically burned politically subversive books. 'Dissenters' from the established Church were denounced, ridiculed, and increasingly persecuted. Supporters of the crown were promoted; it was finally possible, for example, to release the Earl of Danby, the king's former manager of parliament in the royal interest. King Charles displayed the divine majesty of kingship by touching some 6,000 people for the 'King's Evil,' almost 50 per cent higher than the average total for his reign.[37]

In the winter of 1683/4, in short, the Stuart king's enemies were in exile, prison, or dead, its friends protected and promoted, its patronage networks and political influence expanding daily, and its ideological line toed in pulpits, print, and manuscript communication across the country.[38] England's coldest-ever winter occurred during what might well have been the high-water mark of Stuart absolutism.

This royalist political momentum had various enemies and opponents, but a brutally cold winter does not appear to have been amongst them. First, there was no indication that severe cold was a politically sensitive topic. Just a few months after books by John Milton and Thomas Hobbes were burned in the quad of the Bodleian Library, the hard frost was widely and freely discussed and represented. The Thames froze so hard as to support people, booths, animals, coaches, and indeed an entire encampment of commercial and recreational activities generally described as a 'frost fair.' This was discussed publically in broadside ballads and privately in personal correspondence. It was frequently said that thousands of Londoners attended the fair to gawk, shop, play, or make money. Any trinket was seemingly saleable upon the river, as long as it somehow recorded the oddity of the circumstances: a seven-inch-long spoon, for example, was engraved with 'This Spoon was bought upon the Frozen Thames ⋆ January: 28:1683/4.'[39] Others bought prints or coins, as well as food and drink. Among the guests were King Charles II himself and the royal family.

The hard frost was discussed in newsletters, diaries, memoirs, and official correspondence as an interesting, relevant, or noteworthy event but not as a significant part of a political narrative.[40] It is mentioned, for example, in seventeen newsletters in the collection of Sir Richard Newdigate during the first six weeks of 1684.[41] These letters were privately circulated manuscripts and yet were also created specifically as public documents by authors in the office of the secretary of state. They might therefore be assumed to reflect the regime's positions, and indeed they quite clearly endorse the king's allies and excoriate his enemies, though without indulging in the propagandistic rhetoric of many political pamphlets. There is little hint in these letters that the regime worried about how the cold might be construed by a restless public, and the only explicit commentary on the meaning and significance of the freeze is a mild disdain for the frivolous fun Londoners were finding on the frozen river. On 1 January, for example, the newsletter author wrote:

> the frost season continues that on the Thames some thousands of people walk in a beaten path from near the Bridge, though few attempt to foot it across, being forewarned by the loss of diverse lads who perished in that undertaking. Several poor watermen have erected ice booths and stalls in nature of a fair where the people flock through . . . vanity.'[42]

The 'vanity' of those enjoying the frost fair is occasionally denounced, but is also described in unnecessary detail. These letters otherwise are brief lists of events both domestic and foreign, generally unadorned by lengthy descriptions. And yet they find space to tell readers how a man built a three-story house on the ice and slept in it overnight, all to win a wager, as well as how a bull was baited, a fox hunted, and an ox roasted.[43] Such 'frolics' are juxtaposed to some suffering, such as people falling into the water and drowning, coaches overturning on the ice, and of course the watermen who were unemployed as long as the Thames remained impassable.[44] These letters neither celebrate nor suppress the frost fair, but rather treat it as something worthy of commentary, interesting to readers as a curiosity. In these descriptions there is sometimes a position taken or judgment implied, as in the use of words like 'vanity' and 'frolics', but this contrasts with the unmistakable political agenda behind the description of Whigs and dissenters, described as 'the discontented party,' 'persons of discord,' or 'idle and desolate persons.'[45] All of this contrasts with the near-silence of the official *London Gazette*, but this might be expected from that organ, since its readers might have been expected to know of conditions within London itself. It did not discuss the frost fair, though it did describe matter-of-factly the problems and dangers attendant on frozen harbors and displayed buoys.[46]

Beyond the newsletters, individual allies of the regime in 1684 were also open about describing the depth of the cold and paralyzing effects of the frost. Sir John Reresby was a Tory gentleman in London that winter, working to further royal policies of suppressing Dissenters and securing the surrender of York's corporate charter. In his *Memoirs* he pauses briefly from these events to describe the frost, again listing bull-baiting and coaches running over the ice. He noted that the coasts were frozen 'so that we could have no commerce from any part of the world, no ships or boats being able either to go out or come in.'[47] This is clearly a problem, but it is not raised as a danger to the state nor as a serious threat to its people. There is no sense of connection between this vignette and the political chronicle in which it is embedded, no causal relationship asserted or implied between the extreme cold and the political work that kept Reresby busy that winter.

The king's younger brother James, Duke of York, whose Catholicism was the primary problem around which the Whig opposition coalesced, also raised the frost in a context that was resolutely apolitical. James maintained correspondence with Prince William of Orange, the champion of Dutch Protestantism whose invasion in 1688 would topple James from his throne in the 'Glorious Revolution.' As both his uncle and father-in-law, James wrote frequently to William, maintaining superficially friendly relations despite their divergent political interests. His letters of 1683/4, therefore, are notable for a willingness to discuss the severe winter alongside an unwillingness to discuss much of anything else. On 15 January, James wrote to express sorrow that his daughter Mary had suffered from sore eyes, noting that few people had escaped the effects of the sharp winter.[48] His own wife had been let blood, and no fewer than four English peers had recently died and more should be expected should the freeze continue. Before closing, James noted that the Zuider Zee was frozen, responding with a brief description of the coaches and ice skating on the frozen Thames. Then James ended the letter. The newsletter written that same day described a crackdown on Dissenters in London, the City of Oxford's surrender of its charter, a new fleet intended to sail in the spring, and another alleged plotter arrested.[49] None of this, nor any other discussion of politics, had a place in James's letter to William, which restricted itself entirely to the health of family members and acquaintances and the unusual weather. Winter weather, in James's letters to William of 1683/4 as well as in subsequent years, seems to have been considered appropriately apolitical to discuss in polite correspondence with a man who was simultaneously family and a serious political threat.[50]

Roger Morrice was, somewhat like Reresby, both a keen observer and also a participant in politics, but his commitments were on the opposite side of the Whig/Tory divide. As a dissenting minister and an active agent of dissenting political leaders, the winter of 1683/4 was a dangerous period for Morrice's allies and, so he must have often feared, for himself. His enormous 'entering book' records his wide knowledge of politics and political affairs and may in fact have been the basis for manuscript newsletters similar to those received by Newdigate. It is significant, therefore, that Morrice recorded the progress of the winter frost and the fair on the Thames in terms very similar to the letters produced by the crown's own agents. He did not suggest that it might lead to rebellion, revolt, disorder, or governmental weakness, nor that anyone else would hope or fear that it might do so. For Morrice, as for the Tory newsletter author, the Tory historian Reresby, and for the Duke of York himself, England's coldest winter appears as an apolitical interlude, entirely separate from and unconnected to political news.[51]

These more or less public documents discuss the hard frost in a mode very similar to that expressed in private diaries. The Oxford antiquary Anthony Wood described the progress of the weather from December to February, noting when it thawed and when it was 'so cold, as not the like was known by man.'[52] At the beginning of this list of freezes and thaws, Wood notes that 'a child or two' were frozen near Oxford. At its end he summarizes that the weather 'Did a great deal of mischief. In gardens killed laurels, bays . . . hedges.' Such concise entries were characteristic of Wood, who was not very prone to enlarging on the political implications of such events. John Evelyn's diary, however, did regularly pursue such considerations. The introduction described his response to the Restoration and Popish Plot, and he also described the Great Fire of 1666 as a 'most dismal judgment' that punished the sins of subjects and monarch alike: 'our prodigious ingratitude, burning lusts, dissolute court, profane and abominable lives.'[53] This is exactly the sort of politically destabilizing reading of disaster in which, as Parker argues, early modern people regularly engaged. The winter of 1683/4, however, is recorded as a wonder and a source of private suffering but not as a judgment on the state nor as a possible contributor to dangerous political divisions.

Evelyn describes the frost fair in some detail, recording, like Newdigate's newsletters and like many cheap pamphlets and ballads, the booths, games, food and drink for sale, animals tormented, chased, and cooked, the ice skating, carriages and other bustling traffic running over the ice, as well as various shows and entertainments.[54] But unlike some commentators, Evelyn was not blind to the cold's dangers. While it was common to record unusual deaths or accidents (like Wood's frozen boys or the Newdigate letters' mention of fairgoers falling into the freezing water), Evelyn was more sensitive to damage to the poor and to the natural world. As England's leading advocate of re-forestation and gardening, Evelyn was naturally appalled by the toll the frost took on his gardens, where he found the 'greens and rare plants utterly destroyed.'[55] Elsewhere trees appeared 'lightning-struck,' deer parks were 'destroyed,' while 'fowl, fish, and birds, and all our exotic plants and greens, universally perishing.' London's people were hurt by the suspension of seaborne trade as well as the inability of brewers or others requiring water from 'pipes and engines' to perform their work. Fuel became prohibitively expensive for the poor, and yet enough Londoners burned fires to make the city even more polluted than usual.[56] 'London,' Evelyn claimed, 'by reason of the excessive coldness of the air hindering the ascent of the smoke, was so filled with the fuliginous steam of the sea coal, that hardly could one see across the street, and this filling the lungs with its gross particles, exceedingly obstructed the breast, so as one could scarce breathe.' A few days later he recorded again that 'miserable were the wants of the poor.' In sum, for Evelyn, the frost fair was only half, and a rather grotesque half, of the story: 'a bacchanalia, triumph, or carnival on the water, whilst it was a severe judgment on the land.'

Like the authors discussed, Evelyn shows no fear (or hope) that such a judgment might lead to political instability or otherwise aid the cause of the king's enemies. But such a 'severe judgment' was not entirely without significance to the crown. In mid-January, Evelyn dined with Henry Compton, the Bishop of London, and lobbied on behalf of the poor of his suburban parish of Deptford. He hoped to direct some of the £300 in charity that the king had promised to the poor of London, Westminster, and their surrounding suburbs.[57] Evelyn was not alone; others also lobbied on behalf of their own congregations' claims on royal charity or, in the case of one projector who wanted to collect and administer voluntary offerings, on behalf of themselves.[58] The absence or insufficiency of charity during hard winters was a commonplace complaint, and the crown therefore publicized its own contributions.[59] The addition of these funds, such as the £30 received by one London parish, must have helped significantly to maintain some of those most in need, but other parish records indicate that voluntary offerings from parishioners themselves were more significant than royal bounty.[60] The crown's goal was not, however, to end all suffering but to demonstrate its own benevolence and care. The sum of £300 spent over a few weeks was undoubtedly useful but would not have come close to answering the needs of a metropolis with tens of thousands in need, and furthermore appears as a mere token gesture compared to an annual budget of about £2 million.[61] The suffering of the deserving poor was probably the most severe aspect of a much more widespread period of dead trade and depressed production.[62]

The crown, therefore, was aware of the need to be seen acting in response to the needs of the deserving poor, but the resources devoted to this end were not what would have been forthcoming in response to a real political danger. 'A judgment on the land' the winter of 1683/4 may have been, but neither the king nor his brother and heir nor their allies nor their enemies believed that such a judgment would imperil the state. The reverse was more likely, as was demonstrated by a balladeer who argued, in doggerel verse, that God had sent the winter to punish not the king but his enemies.[63]

Deep freezes and political news

If unusually hard winters had been considered as dangerous political moments, we might expect to find them analyzed in the correspondence of diplomats, whose duties required them to be particularly keen observers and analysts of politics. As Noah Millstone has argued, early modern diplomats and other professional connoisseurs of political news were especially concerned with reading the political implications of events.[64] The practice of statecraft was informed by readings of Tacitus and Machiavelli that emphasized the importance of searching into the secret meanings of political decisions and behaviors. 'Seeing like a statesman' meant, for such men, assuming that the true political significance of events required expert analysis, as kings and ministers pursued reason of state through misdirection, dissembling, and subterfuge. They sought any advantage and were keen to discern any enemy weakness.

A report from the Venetian ambassador in January of 1608 was typical of this approach. It examined the significance of negotiations between the Spanish and the Dutch, discussing what influence England might wish to retain in the Netherlands, how King James's prudence might affect Spanish influence over England, and how all of this might matter to the French. It explored why the Irish rebel Earl of Tyrone had recently been called to Brussels, concluding that the Spanish intended to use him to play a double game, simultaneously threatening and wooing James. Having carefully construed this web of interconnected diplomatic developments, the relation then moved on to a few discreet topics. First, the English court's current pre-occupation with producing a masque, then a dispute regarding a ship whose cargo could not be unloaded because of the thick ice on the Thames. The ambassador's following letter similarly discussed international affairs, similarly interpreting a range of political news. He then described the cold, which was 'intenser than any memory of man.' This was only worth noting because it had delayed communications with Europe, so the situation in the Netherlands was less clear than it should have been.[65]

For English diplomats writing at the same time, the hard winter of January 1608 was similarly noteworthy but also similarly without political significance. George Carew wrote from Paris to Thomas Edmondes in Brussels that 'great snows' and 'continual frosts' had imprisoned Henri IV in Paris, but he offered no suggestion that this had any bearing on the other political news in his letter.[66] Henry Wotton in Venice found the cold similarly striking, noting that it had caused in him 'an indisposition of body' which had prevented his sharing news. Yet the entire business of news itself seemed less important under such circumstances, 'insomuch as the affections of curiosity and harkenings after news are turned rather into counsels of necessity and harkening after meat.'[67] The time was full of accidents; gentlemen had fallen on the street and broken their heads, citizens had frozen to death while hunting birds, there were cases of hands frostbitten, 'stupefied and gangrened with the cold,' and a courier and his horse eaten by starving wolves. This was all worth describing, but it bore no relation, for Wotton, on the political affairs that occupied most of his time and attention. From Cologne, a newsletter also recorded 'great snows and unbearable cold over the low country, several men and beasts have been torn and eaten by wolves, and people are dying of the cold.' But again no connection was made between this observation and the letter's other political, diplomatic, and military news.[68]

Finally, from London John Chamberlain described a similarly wondrous cold and the resulting frost fair on the Thames. In the country there was 'a great mortality' of fish, of cattle, birds, and coneys. In the capital fruit sellers, gamblers, and alewives had set up shop on the frozen river itself. 'Certain youths burnt a gallon of wine upon the ice and made all the passengers partakers.' 'But the best,' according to Chamberlain, 'is of an honest woman (they say) that had a great longing to have her husband get her with child upon the Thames.'[69] Chamberlain's tone here is far more lighthearted than Wotton's was, and he presented the frost more as a wonder than as a series of dangers. But Chamberlain agreed with Wotton, Carew, and the Venetian ambassador Zorzi Giustinian that this severe cold had no serious political implications. It hindered the circulation of news but barely qualified as news itself. To comment on the cold was to depart from serious matters of diplomacy and statecraft, a lowering of the tone from business to familiar and unserious talk. These diplomats did not expect that such cold might contribute to any kind of revolt, nor did they suspect that anyone might think it should do so. At the heart of the English regime, Salisbury received frequent reports from Newmarket in December of 1607, many of which mention the harsh weather but none of which suggest that it mattered politically.[70] The only clear response from the crown was an allocation of £200 to be distributed amongst the 'poor watermen' of the Thames who were unable to ferry customers while the river was frozen.[71] The lord mayor and aldermen of London responded to conditions by fixing fuel prices for poor purchasers and by launching an investigation into the management of its subsidized fuel charity.[72]

In some of this political correspondence, intense cold is not only not news, it is the opposite of news, a hinderer and inhibiter of the usual circulation of politically significant communication. There were multiple ways, however, for deep cold to affect communications and even military strategy. Statesmen did not worry excessively about the shivering poor, but they did consider how frozen harbors and highways might alter the movements of shipping, letters, trade goods, and troops. During the 1660s and 1670s, Secretary of State Joseph Williamson was frequently informed, presumably following his own instructions, of winds and weather conditions along England's eastern coast. His correspondents often separated this information from other discussions that were clearly of greater political moment. In one typical case, a letter of 1668 first described a radical preacher who had departed Harwich for the continent and then moved on to the arrival of foreign university students. 'I have no more,' it concluded, but news of the passage of the London coal fleet, an overnight frost, and the early spring blooming of gooseberries and apple trees.[73] Hard freezes changed travel times, which could matter greatly during sensitive missions. In 1532, Thomas Cromwell was informed by an agent of 'perilous' travel conditions that threatened to delay Thomas Cranmer's return to England, where he was to be installed as Archbishop of Canterbury.[74]

In other cases, freezing conditions were militarily beneficial, at least to one side in a conflict. The severe conditions of 1708/9 prompted suffering across western Europe but with particular significance in France, where Britain and its allies had stretched the military and financial capacities of Louis XIV's state. Bishop Burnet's *History of My Own Time*, which says nothing at all about weather during its discussion of the political events of 1683/4, argues that the 'violent Frost, which continued the longest of any in the Memory of Man' was a providential boon to Britain and its allies. He described 'such a degree of misery in France that many died in the extremity of cold and hunger' and a subsequent breakdown of order.[75] Modern historians agree that in 1708–1709, France experienced not only a very difficult winter but also disastrous harvests, leading to famine, social instability, tax revolts, and a consequent sputtering of its war machine that led directly to peace negotiations.[76] This moment of danger for France, however, was a boon for Britain, whose strong military position helped it to extract valuable concessions during the peace treaties concluded at Utrecht in 1713. The broader point that conditions which hurt one side in a war are correspondingly beneficial for the enemy may be thought obvious and indeed was clear enough to early modern observers who either cursed or blessed winter conditions according to their specific conditions and needs.

Conclusion: hardship and opportunity

This discussion has intended to show, first, that the hardest winters of the Little Ice Age were periods when political stability was possible and even more that they might, as in the case of 1683/4, be periods of monarchical strength and royalist initiative very uncommon during the seventeenth century. More important than any objective judgment of the political situation during these moments, however, is how they were perceived by contemporaries, whether early modern people suspected that instability and rebellion were especially likely during these periods. The claim made by Geoffrey Parker and some other historians, that climate change led states to totter, turns in part on this association, on the belief that unusually bad weather led subjects to suspect that their monarchs lacked legitimacy and were being punished by heaven.[77] Perhaps there were such destabilizing readings of natural processes somewhere in the early modern world, but the evidence from England does not support the claim that climatic events were read as evidence of the Stuart monarchs' deficiencies.

During the hard winter of 1607/8, some of England's most sophisticated and Machiavellian-minded political actors assessed freezing conditions with no reference to politics. Freezing rivers and frostbitten digits were mentioned in letters as a break from rather than a contribution to political news. Hard frosts, for these and other correspondents throughout the early modern period, were often anti-news, conditions that hindered the circulation of political information but themselves had no political meaning unless they threatened (or promised) to alter the conduct of warfare. In 1683/4, English observers and political actors of all political persuasions found the hard frost similarly a-political. A Whig like Roger Morrice and a Tory like John Reresby offered similar stories of that winter as a curiosity with no real significance. At a time when his political interests prospered, the strange weather was the one safe topic that the Duke of York could discuss in correspondence with his son-in-law (and future deposer), William of Orange.

Across the century, such moments were seized upon as opportunities for monarchs and lesser governors to display their beneficence and paternal care. Parker's and Cressy's discussion of the reign of Charles I often imply that cold weather contributed to bad harvests and subsistence crises and that it was natural for starving people to blame their rulers during hard times. The public distribution of charity during hard winters, however, shows that this is misguided for two reasons. First, hard winters could be read as entirely lacking providential significance because they were slight variations on a very normal pattern. Winter was always cold, so an unusually cold winter was not the kind of unique and dramatic divergence from common patterns that truly alarmed contemporaries. Second, freezes allowed monarchs and magistrates to demonstrate benevolence. Decades of work by social historians have shown that this is also true of dearths; bad harvests and high prices were more likely to enhance state power than to lead to uncontainable violence.[78] Hungry villagers realized that strong states were useful tools in crisis conditions, and hunger therefore led, in many cases, to popular participation in the process of government rather than to riotous destruction.

Rulers too knew that periods of suffering offered them opportunities, anticipating Rahm Emmanuel's dictum that 'You never want a serious crisis to go to waste' . . . it's an opportunity to do things that you think you could not do before.'[79] Against Parker's view that disasters were inevitably viewed as providential judgments against illegitimate rulers, one could cite a venerable tradition of rulers using disasters to buttress their image as benevolent or effective governors. Joseph and Pharaoh during the famine in Egypt, Titus after the eruption of Vesuvius, the Marquis of Pombal after the Lisbon earthquake, Raphael Trujillo after the San Zenón Hurricane in the Dominican Republic, and President Obama after Hurricane Sandy on the Atlantic coast of the United States: such examples serve as reminders that disasters offer opportunities as well as dangers.[80] Parker makes a similar observation but argues that the shock of the 1640s led to gradually adopted 'coping strategies,' encompassing a huge range of behaviors observable during the century after 1650.[81] But the English response to severe winters is characterized by continuity across the century of this 'crisis'; both the winters of 1607/8 and 1683/4 were met with token charity from those in power and a collective shrug of the shoulders regarding weather's impact on high politics. Given the ways that such winters offered at once hardship and political opportunities, perhaps the question to ask about the Little Ice Age in Britain is not how it caused revolts but rather how its difficult conditions could have allowed for (or even prompted?) state formation, economic development, and imperial expansion.[82] Answering this will require attention not only to the dynamic of an increasingly connected world but also to the agendas that prompted early modern people to endure, collaborate with, and seek to appropriate the powers of the state.

Notes

1 *The Diary of John Evelyn, Vol. 4: Kalendarium, 1673–1689*, ed. E.S. de Beer (Oxford: Oxford University Press, 1955), 359–366, quotations on 360, 365, 362.
2 *The Diary of Samuel Pepys: Vol. 7 1666*, ed. Robert Latham and William Matthews (Berkeley: University of California Press, 1974), 29; Steven Pincus, 'John Evelyn: Revolutionary', in *John Evelyn in His Milieu*, ed. John Hunter and Francis Harris (London: The British Library, 2003), 188.
3 *The Diary of John Evelyn, Vol. 3: Kalendarium, 1650–1672*, ed. E.S. de Beer (Oxford: Oxford University Press, 1955), 450–407; Gillian Darley, *John Evelyn: Living for Ingenuity* (New Haven: Yale University Press, 2006), 154.
4 See the recent synthesis by John L. Brooke, *Climate Change and the Course of Global History: A Rough Journey* (Cambridge: Cambridge University Press, 2014), Chapter 10.
5 Collected in *Crisis in Europe 1560–1660*, ed. Trevor Aston (New York: Doubleday, 1967).
6 The excellent recent *Oxford Handbook to the English Revolution*, for example, only discusses the general crisis twice in thirty-three chapters, one of which describes this approach as 'problematic,' while the other cites only one article written since 1970, authored by Geoffrey Parker (for whom see what follows). *The Oxford Handbook to the English Revolution*, ed. Michael Braddick (Oxford: Oxford University Press, 2015), 79, 579.
7 Jan de Vries, 'The Economic Crisis of the Seventeenth Century after Fifty Years,' *Journal of Interdisciplinary History*, 40, no. 2 (2009): 152. Also *The General Crisis of the Seventeenth Century*, ed. Geoffrey Parker and Lesley M. Smith (London: Routledge, 1978; 2nd ed., 1997); Jack Goldstone, *Revolution and Rebellion in the Early Modern World* (Berkeley: University of California Press, 1981), 25–26; *Early Modern Europe: From Crisis to Stability*, ed. Philip Benedict and Myron P. Gutmann (Dover, DE: University of Delaware Press, 2005).
8 Geoffrey Parker, 'Crisis and Catastrophe: The Global Crisis of the Seventeenth Century Reconsidered', *American Historical Review* 113, no. 4 (2008): 1053–1079; Geoffrey Parker, *Global Crisis: War, Climate Change and Catastrophe in the Seventeenth Century* (New Haven: Yale University Press, 2013).
9 For England, David Cressy has recently suggested that cold, rainy, or wintery weather during the 1630s undermined the regime of Charles I, though without showing evidence that it did so. David Cressy, *Charles I and the People of England* (Oxford: Oxford University Press, 2015), 54–64. Parker's work has been used to place the collapse of the Ming in a 'global perspective,' but this is an appendix to a political narrative that takes almost no notice of climate. Kenneth M. Swope, *The Military Collapse of China's Ming Dynasty, 1618–44* (New York: Routledge, 2014), Chapter 8. A more precise connection between climatic conditions and specific behaviors and choices is Sam White, *The Climate of Rebellion in the Early Modern Ottoman Empire* (Cambridge: Cambridge University Press, 2013).
10 Parker, *Global Crisis*, 17–25, on Japan 484–506.
11 Paul Warde, 'Crisis or Coincidence?', *Past and Present* 228 (2015): 287–301, quotes on 291, 293.
12 Kenneth Pomeranz, 'War, Weather, and Welfare: Persistence and Change in Geoffrey Parker's *Global Crisis*', *Historically Speaking* 14, no. 5 (2013): 30–33.
13 Jan de Vries, 'The Crisis of the Seventeenth Century: The Little Ice Age and the Mystery of the "Great Divergence"', *Journal of Interdisciplinary History* 44, no. 3 (2014): 369–377, especially 374. Regina Grafe's recent survey of early modern European economic history similarly stresses that economic developments were more complex than an over-reliance on climatic trends can explain. Regina Grafe, 'Economic and Social Trends', in *The Oxford Handbook of Early Modern European History, 1350–1750: Vol. 1: Peoples and Place*, ed. Hamish Scott (Oxford: Oxford University Press, 2015), 279–280.
14 In particular Conrad Russell, *Unrevolutionary England: 1603–42* (London: Hambledon, 1990), Chapters 7–10. A recent issue of the *Huntington Library Quarterly* reviewed the contributions and legacies of this scholarship, with a particularly good critique from Peter

Lake. Peter Lake, 'From Revisionist to Royalist History: Or, Was Charles I the First Whig Historian?', *Huntington Library Quarterly* 78, no. 4 (2015): 657–681.

15 Conrad Russell, *The Causes of the English Civil War* (Oxford: Clarendon Press, 1990), 207.

16 A good narrative summary of this approach is Michael Braddick, *God's Fury England's Fire: A New History of England's Civil Wars* (London: Penguin, 2008); see also *The Huntington Library Quarterly* cited in note xiv; Peter Lake, 'Post-Reformation Politics, or on Not Looking for the Long-Term Causes of the English Civil War', in *Oxford Handbook to the English Civil War*, ed. Michael Braddick (Oxford: Oxford University Press, 2015), 21–42.

17 E.P. Thompson, 'The Moral Economy of the English Crowd in the Eighteenth Century', *Past and Present* 50 (1971): 76–136.

18 Such work includes Buchanon Sharp, *In Contempt of All Authority: Rural Artisans and Riot in the West of England, 1586–1660* (Berkeley: University of California Press, 1980); John Walter, *Crowds and Popular Politics in Early Modern England* (Manchester: University of Manchester Press, 2010); Andy Wood, *Riot, Rebellion, and Popular Politics in Early Modern England* (New York: Palgrave Macmillan, 2002); John Bohstedt, *The Politics of Provisions: Food Riots, Moral Economy, and Market Transition in England, c. 1500–1850* (Farnham: Ashgate, 2010).

19 Keith Wrightson, *English Society 1580–1680* (New Brunswick, NJ: Rutgers University Press, 1982), 175, 178.

20 During the late-1640s, according to Hindle, the poor prompted magisterial action at a time when many aspects of local governance had broken down. Steve Hindle, 'Dearth and the English Social Revolution: The Harvest Crisis of 1647–50', *Economic History Review* 61 (2008): 64–98.

21 Andy Wood, *The 1549 Rebellions and the Making of Early Modern England* (Cambridge: Cambridge University Press, 2007); Krista Kesselring, *The Northern Rebellion of 1569: Faith, Politics, and Protest in Elizabethan England* (London: Palgrave Macmillan, 2007); Diarmid MacCulloch and Anthony Fletcher, *Tudor Rebellions* (6th ed., London: Routledge, 2015).

22 Parker, 'Climate and Catastrophe'.

23 Parker, *Global Crisis*, 333.

24 Laura A.M. Stewart, *Rethinking the Scottish Revolution: Covenanted Scotland 1637–51* (Oxford: Oxford University Press, 2016), 45, n.48.

25 Parker, *Global Crisis*, 8–9.

26 Morgan Kelly and Cormac Ó Gráda. 'The Waning of the Little Ice Age: Climate Change in Early Modern Europe', *Journal of Interdisciplinary History* 44, no. 3 (2014): 301–325. But see the forceful and convincing rebuttals in the same issue by Sam White, 'The Real Little Ice Age', *Journal of Interdisciplinary History* 44, no. 3 (2014): 327–352, and Ulf Büntgen and Lena Hellmann, 'The Little Ice Age in Scientific Perspective: Cold Spells and Caveats', *Journal of Interdisciplinary History* 44, no. 3 (2014): 353–368.

27 Wolfgang Behringer, 'Weather, Hunger, Fear: Origins of the European Witch Hunts in Climate, Society, and Mentality', *German History* 13 (1995): 1–27; Wolfgang Behringer, *A Cultural History of Climate*, trans. Patrick Camiller (Cambridge: Polity Press, 2010); Christian Pfister, 'Climatic Extremes, Recurrent Crises and Witch Hunts: Strategies of European Societies in Coping with Exogenous Shocks in the Late Sixteenth and Early Seventeenth Centuries', *Medieval History Journal* 10 (2006): 33–73.

28 The importance of the 'peccatogenic worldview' was further underscored by Parker during the discussion of *Global Crisis* at the Social and Economic History of the Early Modern World, 1500–1800, Seminar, Institute of Historical Research, University of London, 2 May, 2014.

29 Alexandra Walsham, *Providence in Early Modern England* (Oxford: Oxford University Press, 1999).

30 Reingard Esser, 'Fear of Water and Floods in the Low Countries', in *Fear in Early Modern Society*, ed. William G. Naphy and Penny Roberts (Manchester: Manchester University Press, 1997), 62–77; Elaine Fulton, 'Acts of God: The Confessionalization of Disaster Reformation Europe', in *Historical Disasters in Context: Science, Religion, and Politics*, ed. Andrea Janku, Gerrit J. Schenk and Franz Mauelshagen (Abingdon: Routledge, 2012), 54–74.

31 In addition the works cited earlier, see Franz Mauelshagen, 'Disaster and Political Culture in Germany since 1500', in *Natural Disasters, Cultural Responses: Case Studies toward a Global Environmental History*, ed. Christof Mauch and Christian Pfister (Plymouth: Lexington, 2009), 41–75; Elaine Fulton and Penny Roberts, 'The Wrath of God: Explanations of Crisis and Natural Disaster in Pre-Modern Europe', in *History at the End of the World: History, Climate Change, and the Possibility of Closure*, ed. Mark Levene, Rob Johnson and Penny Roberts (Penrith: Humanities-E Books, 2010), 63–79; John Morgan, 'Understanding Flooding in Early Modern England', *Journal of Historical Geography* 50 (2015): 37–50.

32 The topic is nicely surveyed in Christian Pfister, 'Weather, Climate, and the Environment', in *Oxford Handbook of Early Modern European History, 1350–1750: Vol. 1, Peoples and Place*, ed. Hamish Scott (Oxford: Oxford University Press, 2015), 70–93.

33 Parker, *Global Crisis*, 17.

34 Anecdotal evidence suggests that parts of southern England must have been very much colder than this during the depths of the frost, from late December to early February. The register of one parish in Wiltshire, for example, recorded that the ground was frozen to a depth of three feet. Leigh parish register, in *Gleanings from Wiltshire Parish Registers*, ed. Steven Hobbs (Chippenham: Wiltshire Record Society, 2010), 142. Thanks to John Morgan for this reference.

35 MET Office, 'Seasonal Central England Temperature, 1659 to 2016'. http://www.metoffice.gov.uk/hadobs/hadcet/ssn_HadCET_mean.txt. Accessed 12 August 2016.

36 This literature is surveyed in Alvin Snider, 'Hard Frost, 1684', *Journal for Early Modern Cultural Studies* 8, no. 2 (2008): 8–32.

37 Stephen Brogan, *The Royal Touch in Early Modern England: Politics, Medicine, and Sin* (Woodbridge: Boydell for the Royal Historical Society, 2015), 101.

38 This period is surveyed in Gary S. De Grey, *Restoration and Revolution in Britain: Political Culture in the Era of Charles II and the Glorious Revolution* (London: Palgrave Macmillan, 2007), 202–210; Tim Harris, *Restoration: Charles II and His Kingdoms, 1660–1685* (London: Penguin, 2005), Chapter 5; Grant Tapsell, *The Person Rule of Charles II, 1681–85* (Woodbridge: Boydell, 2007).

39 Lawrences Auctioneers, 'Old Spoon gets its Just Deserts in Winter's Tale'. http://www.lawrences.co.uk/news.php?m=1&y=2016. Accessed 12 March 2016.

40 In addition to the discussion that follows, the frost in general, and often London's frost fair in particular, was commented upon by Anne, Dowager Countess of Manchester, Northamptonshire Record Office, FH 4391, Anne, Dowager Countess of Manchester, to Lady Hatton, 15 January, 1683/4; Sir John Lowther of Stockbridge to Sir Daniel Fleming, 26 January, 1683/4, *HMC Twelfth Report, Appendix, Part VII. The Manuscripts of S.H. Le Fleming, Esq. of Rydal Hall* (1890), 193; *The Letters of William Freeman, London Merchant, 1678–1685*, ed. David Hancock (London: London Record Society, 2002), 363.

41 Folger Shakespeare Library (hereafter FSL), W.C. 1472–96. These were consulted online at the Henry E. Huntington Library, January 2016.

42 FSL W.C. 1472

43 FSL W.C. 1483, 1485, 1491, 1494.

44 FSL W.C. 1491.

45 FSL W.C. 1464, 1466, 1481.

46 *London Gazette*, Issue 1899, 28–31 Jan. 1684; Issue 1900, 31 Jan.–4 Feb. 1684; Issue 1903, 11–14 Feb. 1684.

47 John Reresby, *Memoirs of John Reresby: The Complete Text and a Selection from His Letters*, ed. Andrew Browning (London: Royal Historical Society, 1991), 328–329.

48 *Calendar of State Papers Domestic* [Hereafter *CSPD*] *1683–84*, 216.

49 FSL L.C. 1480.

50 Subsequent discussions of winter cold *CSPD 1683–84*, 263, 281; *CSPD 1686–87*, 123, 139, 324, 329, 345, 400; *CSPD 1687–89*, 104.

51 A similar approach is taken in Narcissus Luttrell, *A Brief Historical Relation of State Affairs from September 1678 to April 1714*, 6 vols. (Oxford: Oxford University Press, 1857), i, 294–299.

52 *The Life and Times of Anthony Wood, Antiquary of Oxford, 1632–1695, Described by Himself*, ed. Andrew Clark, 5 vols. (Oxford: Clarendon Press for the Oxford Historical Society, 1894), iii, 88.
53 *Kalendarium, 1650–72*, ed. De Beer, 464.
54 This and the following discussion of Evelyn uses his 'Diary' entries from 27 December 1683–8 February 1684, *Kalendarium, 1673–1689*, ed. De Beer, 357–366.
55 The loss of trees and plants is also stressed in an anonymous description of the frost British Library (hereafter BL), Harley MS 6992, f. 15v; and Sir Robert Southwell's letter to Evelyn from Bristol of 8 January, *Diary of the Times of Charles the Second by the Honourable Henry Sidney, (Afterwards Earl of Romney) Including His Correspondence with the Countess of Sunderland, and Other Distinguished Persons at the English Court: To Which Are Added, Letters Illustrative of the Times of James II and William III*, ed. W. Blencowe, 2 vols. (London: Henry Colburn, 1843), ii, 243.
56 For Evelyn's concern with air pollution, see William M. Cavert, *The Smoke of London: Energy and Environment in the Early Modern City* (Cambridge: University of Cambridge Press, 2016).
57 King Charles to the Bishop of London and to the Dean of Westminster, 4 January, 1684, The National Archives of the United Kingdom [hereafter TNA] SP 44/53, f. 127, 129. £100 each was paid to the Bishop of London, the churchwardens of St. Margaret's Westminster, and Stepney (home of many sailors and watermen), *Moneys . . . for Secret Services of Charles II and James II*, vol. 53 (London: Camden Society, 1851), 81–82; also TNA PC 2/70, 110.
58 TNA SP 29/436/68, Dr. Bréval to Sir Joseph Williamson, 26 January, 1684; TNA SP 44/55, f. 306.
59 On the coldness of charity during hard winters see, for example, Thomas Tryon, *Modest Observations on the Present Extraordinary Frost* (London: George Larkin, 1684).
60 For £30 received from the Bishop of London during the 'late great frost,' London Metropolitan Archives [hereafter LMA] P69/BOT4/B/009/MS04525/007B, St Botolph without Bishopsgate churchwardens accounts, f. 25. For other London parishioners recording primarily or only voluntary contributions from parishioners, LMA P69/JNB/B/006/MS00577/002, St. John the Baptist, Walbrook, overseers' accounts 1679–1728, f. 33; P69/BOT2/B/018/MS02626, St Botolph Aldgate Churchwardens' accounts 1683–99, f. 1v; LMA DRO/052/133, St Mary Acton (Ealing) churchwardens accounts, 1674–99, p. 100.
61 Narcissus Luttrell claimed that the King gave £2000, but I have found no confirmation of this figure. Luttrell, *Brief Historical Relation*, i, 296.
62 The frost was blamed for sugar being 'a dull commodity,' *Letters of William Freeman*, 490. Narcissus Luttrell claimed that the prices of 'all commodities rose,' Luttrell, *Brief Historical Relation*, 294. For the steep decline of brewing TNA CUST 48/6.
63 *The Whigs Hard Heart for the Cause of the Hard Frost* (London, 1684).
64 Noah Millstone, 'Seeing Like a Statesman in Early Stuart England", *Past and Present* 223 (2014): 77–127.
65 *Calendar of State Papers Relating to English Affairs in the Archives of Venice, 1607–1610*, ed. Horatio F. Brown (London: HMSO, 1904), 81–83.
66 BL Stowe 169, Edmondes Papers Vol. IV, 1607–8, f. 240.
67 Ibid., f. 269. Also *The Life and Letters of Henry Wotton*, ed. Logan Pearsall Smith, 2 vols. (Oxford: Clarendon Press), i, 409.
68 The Cecil Papers [CP] 194/113, (calendar reference Vol. 20, p. 15) and CP 194/120 (Vol. 20, 62).
69 BL Stowe 169, John Chamberlain to Dudley Carleton, 19 Jan. 1607/8, f. 247v.
70 CP 123/93, 123/94, 123/150, 108/17 (Vol. 19, 360, 361, 362, 368).
71 *CSPD 1603–10*, 400.
72 LMA Rep. 28, f. 146; LMA COL/AD/1/28 Letter book CC 1, f. 272-v.
73 TNA SP 29/236/14, Silas Taylor to Williamson, Harwich, Mar 5 1667/8, f. 28. Other informants describing frosts during the 1670s, both in times of war and peace, include

TNA SP 29/333/102 f.154, SP 29/332/, f. 259; SP 29/375/78, f. 166; SP 29/387 f. 126; SP 29/398, f. 153.

74 TNA SP 1/72, 9 Dec. 1532. R.O.S. [Stephen] Vaughan to Cromwell, f. 121. Also TNA SP 1/72, 19 Dec 1532, John Bunolt to Thomas Cromwell, f. 141.

75 Gilbert Burnet, *Bishop Burnet's History of His Own Time: Vol. 2 from the Revolution to the Conclusion of the Treaty of Utrecht, in the Reign of Queen Anne* (London: Thomas Ward, 1734), 511–512; *A Supplement to Burnet's History of My Own Time*, ed. H.C. Foxcroft (Oxford: Clarnedon, 1902), 420.

76 W. Gregory Monahan, *Year of Sorrows: The Great Famine of 1709 in Lyon* (Columbus: Ohio State University Press, 1993); Marcel Lachiver, *Les Années de misère: la famine au temps du Grand Roi, 1680–1720* (Paris: Fayard, 1991); John A. Lynn, *The Wars of Louis XIV 1667–1714* (New York: Longmann, 1999), 325, 336.

77 It should be repeated that the structural and economic aspects of Parker's argument, which are discussed previously in this chapter, are not under consideration here.

78 Besides the works cited in note 11, see also Paul Slack, 'Books of Orders: The Making of English Social Policy, 1577–1631', *Transaction of the Royal Historical Society* 5th series 30 (1980): 1–22; Ian Archer, *The Pursuit of Stability: Social Relations in Elizabethan London* (Cambridge: Cambridge University Press, 1991).

79 Speech to the Wall Street Journal CEO Council, November 20, 2008, (podcast.mktw.net/wsj/audio/20081120/pod-wsjemmanuel/pod-wsjemmanuel.mp3) Accessed 26 February 2016.

80 Genesis 47:13–26; Gaius Suetonius Tranquillus, *The Twelve Caesars*, trans. Robert Graves (London: Penguin, 1989), 296; Kenneth Maxwell, *Pombal: Paradox of the Enlightenment* (Cambridge: Cambridge University Press, 1995); Stuart B. Schwartz, *Sea of Storms: A History of Hurricanes in the Greater Caribbean from Columbus to Katrina* (Princeton, NJ: Princeton University Press, 2015); 243–251, 333–336; Naomi Klein, *The Shock Doctrine: The Rise of Disaster Capitalism* (New York: Picador, 2007).

81 Parker, *Global Crisis*, 590.

82 Some attempts to wrestle with these processes include Dagomar Degroot, 'Never Such Weather Known in These Seas: Climatic Fluctuations and the Anglo-Dutch Wars of the Seventeenth Century, 1652–1674', *Environment and History* 10, no. 2 (2014): 239–273.

7 "A considerable change of climate"

Glacial retreat and British policy in the early-nineteenth-century Arctic*

Anya Zilberstein

Returning to England from hunting whales in the northern seas at the end of summer 1817, William Scoresby, Jr. reported some extraordinary news. Six thousand square miles of the Arctic Ocean, particularly the coasts of the Greenland Sea, had been "perfectly void of ice". Scoresby also documented rising seawater temperatures and unusually large ice floes drifting from the coasts of Iceland and Spitsbergen to areas much farther to the south. Finding what appeared to be a disintegrating ice shelf and a remarkably ice-free Arctic Ocean utterly contradicted what Scoresby – a seasoned whaler and a respected naturalist – thought he knew about the region's climate and the nature of its frozen seascapes. Should these conditions last beyond a single season and the northernmost reaches of the Arctic Ocean also prove to be navigable, it held out the tantalizing possibility of finally resolving "the mystery attached to the existence of a north west passage". But actual endeavours to locate the fabled Northwest Passage, he cautioned, should only move forward if sustained research could show that this situation was not merely anomalous but a historic precedent that could be documented over the course of several return visits. "If such a passage does exist", Scoresby wrote in a letter to his patron, President of the Royal Society Sir Joseph Banks, "it will be found only at intervals of some years; this I deduce from attentive observation of the nature, drift, and general outline of the polar ice".[1]

Despite Scoresby's advice to sponsor more careful research, Banks reacted swiftly. He too was unsure if the Arctic's ice sheets would continue to melt; and he remained sceptical about whether there was a persuasive scientific explanation to account for these unexpected conditions. After centuries of failed attempts to locate the Northwest Passage, contemporary scientific authority and common wisdom suggested that it was close to impossible. Impenetrable ice engulfed the Arctic Circle, George-Louis Leclerc, comte de Buffon, had asserted in the 1770s. In the same decade, British commanders charged with exploring a route through the ice came to a less absolute but still pessimistic conclusion. A route "most certainly exists", they submitted, "but will never be found practicable for mercantile purposes". More recently, Scoresby had himself dismissed the idea. As he declared in the paper "On the Greenland or Polar Ice", which he presented to an Edinburgh scientific society in 1815, an ice-free Arctic was "too improbable to render it necessary to hazard any opinion concerning it". Nevertheless,

Banks felt that such statements no longer held true. "The more I have considered the facts stated in it", he wrote to Scoresby about his 1817 report on the Arctic, "the more I am convinced that the information given in it to the Public for the first time is likely to lead to results highly advantageous to maritime science". Banks made immediate moves to direct what he considered a "matter of Great importance to the Prosperity of this Countrey": discovering a northern maritime route to Asia would allow Britain to claim possession of the coasts of an ostensibly ice-free Arctic Ocean. He urged the Admiralty to devote funds to produce maps of the Far North based on fresh surveys and helped to renew a 1745 Act of Parliament offering bounties for private exploration of polar regions. In 1818, the Royal Navy sent out two Arctic expeditions as a result of Banks's initiatives. These were to be the opening gambits in a series of annual government-sponsored forays in search of a Northwest Passage that notoriously misspent funds and jeopardized or lost lives, most famously in the catastrophic John Franklin expedition of 1845.[2]

This chapter offers a retrospective analysis of Banks's impulsive push to reenergize Arctic exploration in the early nineteenth century by examining how a range of long-established and on-going scientific inquiries into climate change informed his actions. Banks was the chief promoter of science in Britain and its massive empire in the crucial period of state formation between the reconfiguration of empire during the Atlantic revolutions of the 1780s and the end of the Napoleonic Wars. As John Gascoigne and Richard Drayton have carefully demonstrated, it is difficult to overestimate his commanding influence in shaping British investment in science, particularly in funding research subjects that required fieldwork in places far removed from metropolitan institutions. Banks's power in scientific networks derived most obviously from his long tenure as head of the Royal Society from the 1780s until his death in 1820 as well as his close contact with George III. Banks further enhanced these privileged positions by becoming a founding member of or principal advisor to numerous other government agencies and extra governmental learned societies such as the Admiralty, the Colonial Office, the Home Office, the Board of Agriculture, the Royal Gardens at Kew, and the Royal Institution, to name only a few.[3]

Consequently, appreciating Banks's role in linking Scoresby's putative discovery of glacial retreat to British policy in the Arctic offers some insight into the broader historical questions that *Governing the Environment in the Early Modern World* seeks to understand, such as when, why, and how states have turned their attention to climate change in the past. In this case, the simple and perhaps unsurprising answer – nevertheless one that has particularly clear resonances with twenty-first-century controversies over Arctic climate change and sovereignty – is that Banks was willing to act quickly on the implications of a report about climate change that strategically aligned with Britain's long-established commercial, military, and political goals, even though he knew the evidence was extremely flimsy. By demonstrating that the politicization of climate change science is hardly a new phenomenon, this case helps work against the presumption that the political problems of climate change in the Anthropocene are uniquely postmodern.[4]

Moreover, Banks's reaction to Scoresby's sightings in 1817 set off what I propose was likely the first instance, at least in Britain, of metropolitan government sponsorship of scientific activity in direct response to climatic instability. As will become clear, this was a relatively abrupt about-face in Banks's attitude to theories about the dynamics of climate, about which, both as a naturalist and a scientific patron, he had hitherto been largely indifferent, despite the profusion of communications about local climate change that he received directly from British naturalists, both at home and abroad in colonies or former colonies. Although discovering the Northwest Passage before other European nations and dominating the Circumpolar North undoubtedly motivated Banks's reaction, this episode also provides historians with important clues about his opinions in regard to contemporary climate change theory. Before the late eighteenth century, most of his commentary on the topic of climate suggests that he imagined it as a stable phenomenon organized in terms of a neat geography of polar, temperate, and tropical zones according to which Britain could manage resource extraction in its domestic and colonial economy: wool and wheat in England; timber and furs from British North America; fish from Newfoundland; sugar, rice, and indigo from the West Indies; tea from India. He never endorsed the theory of anthropogenic climate change – a hotly debated idea with an ancient pedigree that was affirmed by many of his contemporaries, from prominent naturalists and philosophers such as Buffon, David Hume, and Benjamin Franklin to numerous more obscure writers. Since the sixteenth century, naturalists in the Americas, British Isles, and Europe had been trying to determine the causes and consequences of a variety of climatic changes they supposed were occurring in their locales. The most prevalent argument in the English (later British) Empire was that economic improvement had destabilized regional climates, changes that seemed particularly conspicuous in rapidly developing colonies in the Northern Hemisphere's temperate zones. By the end of the eighteenth century, these and other conjectures about climate change had become a standard part of scientific correspondence, debate, and land management across the Atlantic world.[5]

Banks was thoroughly aware of such discussions, and yet there is frustratingly little indication of his position in regard to them. With few exceptions, he was unforthcoming on the topic of climatic instability in his publications on natural history as well as in his voluminous official, scientific, and private exchanges with hundreds of correspondents in his vast network of contacts around the globe. Yet these few exceptions indicate that Banks first became curious about climatic instability in the 1790s out of concern that England's summers were increasingly shorter and colder, conditions that seemed to him to prevail through the first decade of the nineteenth century, when he first began to record this concern. The continuation of cooler weather contradicted the many attestations of climate warming he had received in letters from naturalists, particularly in North American colonies of settlement. On the other hand, Scoresby's report provided credible evidence for theories of glacial retreat that seemed to him to more plausibly, if still inconclusively, explain the real causes behind local changes in English weather. From the extant scraps of evidence, Banks was a belated and reluctant adopter of ideas about climate warming, though he was much more convinced that it originated somehow in the internal dynamics and effects of glaciers rather than in the economic activities of human societies. Placing the revival of polar exploration in the context of transatlantic debates about a variety of accounts of climate change in the eighteenth and early nineteenth centuries thus helps to explain why Scoresby's account of the Arctic so impressed Banks.[6]

From the early modern period and through the eighteenth century, imperial European states sustained an avowed interest in gathering knowledge about distant climates, but it was mainly local colonial officials and naturalists who were the most prolific promoters of ideas about climate change. Colonial elites were especially active in reviving ancient ideas about climate amelioration or deterioration as a reflection of different land management practices, usually in ways that triumphantly aggrandized their conquest of indigenous territories. According to this idea, the climate was subject to historical transformation: turning unimproved forests into productive farmlands populated by European settlers would ultimately temper the local climate in regions with harsh winters or summers. Average annual temperatures would rise in cold places, and they would become lower in hot places. In addition, by draining marshes and exposing more soil to the effects of wind and sunlight, colonial settlers and planters believed that they were helping to reduce humidity and, in turn, to reduce settlers' exposure to noxious (or 'miasmatic') air. By this line of reasoning, people had the power to improve their surroundings but they had to work continuously to sustain such environmental transformations, because land that was abandoned or mismanaged would revert to its wild state or else become excessively dry and barren. The desertification of lands surrounding the Mediterranean Sea, where the moderate climate of the ancient world had seemingly yielded to long-term decline after the fall of the Roman Empire, served as a powerful cautionary tale of the twin dangers of political decay and land neglect.[7]

To take just one of countless eighteenth-century pronouncements on the process of climatic tempering as a result of agricultural improvement, James Franklin's "The Philosophical and Political History of the Thirteen United States of America", published in 1784, offered a typical description. The tract's first entry focused on New England and began by listing the disadvantages of the region's climate – namely, that despite its situation at a more southerly latitude than England, its winters were colder and longer and its summers hotter, more humid, and shorter. Over the course of nearly two centuries, however, even these conditions had softened as compared to colonists' first experiences in the early seventeenth century:

> Both the heat and cold are now far more moderate, and the constitution of the air in all respects far better, than our people found it at the first settlement. The clearing away of the woods, and the opening of the ground every where, have, by giving a free passage to the air, carried off those noxious

> vapours, which were so prejudicial to the health of the first inhabitants. The temper of the sky is generally, both in summer and winter, very steady and serene. Two months frequently pass without the appearance of a cloud; and their rains, though they are heavy, do not continue any long time.[8]

Proponents of anthropogenic climate theory also came to believe that shifts in the weather were most radical in environments they had recently conquered and populated with European settlers. Climate change, explained Samuel Williams, was "most of all sensible and apparent in a new country, which is suddenly changing from a state of vast uncultivated wilderness, to that of numerous settlements, and extensive improvements". In 1802, a decade after the establishment of Freetown as the capital of British Sierra Leone, for example, one local official declared that its climate was "believed to be already considerably ameliorated" by colonization. According to Williams, a former Harvard professor of natural philosophy based in Vermont and an elected foreign member of the Palatine Meteorological Society, these were not isolated or incidental changes but rather instances of an undeniable "general and universal" phenomenon with a deep chronology. All inhabitable regions of the world where agriculture could be practiced were subject to the same "operation of causes and effects". "In every cultivated country of which we have any account", wrote Williams, "the climate has been found to grow more mild and temperate". The changing air temperatures in North American and African settlements were only the latest manifestation on a regional scale of a much longer, westward-moving global process of environmental change "from the very first creation of the world until now".[9]

To substantiate their claims, Williams and other naturalists documented such changes in particular locations by collecting evidence from a wide range of sources: classical and modern histories, the Old Testament, and instrumental data recorded using thermometers, barometers, and hygrometers. The idea that climate was an unstable entity subject to historical change was implicit in Genesis, especially in the parable of Noah and the Flood, but the secular notion that human intervention could rehabilitate or ruin the climate was scattered throughout the work of ancient Roman writers such as Pliny, Tacitus, and Strabo, in more recent philosophical tracts, political or natural histories of European provinces or overseas colonies, and travelogues. The latter – such as Linnaeus's disciple Pehr Kalm's diary of his journey through the Middle Colonies and Quebec in the 1740s and 1750s, which other naturalists frequently cited – provided descriptions of contemporary weather conditions but also introduced oral history as a source, usually in the form of quotations from conversations with elderly residents about their memories of weather during their childhood. Learned elites with access to measuring instruments produced empirical evidence of weather and climate history, which they circulated in manuscript or published form. These daily reports of temperature and weather conditions also included significant dates such as the first and last frosts, the commencement of the harvest season, animal migrations, and descriptions of violent storms, hurricanes, earthquakes, comet sightings, and eclipses.[10]

In spite of the quantity and diversity of this evidence, some eighteenth-century naturalists completely rejected the notion that political and natural history influenced one another, especially the possibility that economic activity could affect the climate. In 1767, the Virginian John Mitchell wrote to the Board of Trade that, despite what others had claimed, he was certain that "from 150 years experience of American colonization neither the soil nor climate will admit of any such improvements". Quoting at length from Mitchell's dismissive view of climate amelioration, in the 1780s Jamaican planter and naturalist Edward Long lampooned the "vague and unfounded assertion" that northern North America's climate had become tempered by population increase and agricultural improvement. He ridiculed Joseph Priestley's proposition that climate warming could occur as a result of an increased amount of phlogiston (a supposed combustible substance) in the air:

> We shall probably be told, that so far as regards the *physical* maladies of climate, we are to *hope* that in *process of time*, when the atmosphere of these regions shall be more impregnated with phlogistic particles from myriads of reeking dunghills, from the fumes of furnaces, from the fires and smoke of ten thousand crowded cities *hereafter to be built*, and by a general subjection of the soil to agriculture, carried on to the Arctic Circle, they may be considerably alleviated.

Long's image of a temperate, cultivated Arctic landscape crystallized his view of the absurdity of this hypothesis.[11]

Other naturalists were more equivocal. Many claimed either that they had themselves experienced an unstable climate or felt compelled to account for fossils, instrumental records, and other material evidence of climatic changes in the near or distant past. Uncertain about whether the causes of seemingly remarkable meteorological phenomena were anthropogenic in origin, some linked strange weather to disruptive but temporary seismic events such as the 1755 Lisbon earthquake or to the floating black particles of ash emitted from the eruption of Iceland's Laki volcano in 1783, which darkened the skies in Britain and continental Europe. John Mervin Nooth, a naturalist in Quebec, attributed the increasingly warmer temperatures he documented in his journal of daily thermometer readings as well as anecdotal information gleaned from the province's "oldest inhabitants" to the supposed eruption of an obscure volcano located somewhere in the Saint Lawrence River Valley.[12]

What no one knew for certain – including those who accepted the possibility of anthropogenic causes – was whether colder or warmer seasonal conditions were signs of permanent climate change. Benjamin Franklin was open to "the Truth of the common Opinion, that the Winters in America are grown milder". "But whether enough of the Country is yet cleared to produce any sensible Effect, may yet be a Question", he wrote to a colleague at Yale College in 1763. Moreover, Franklin wondered if milder weather might be an indication of superficial variability and therefore decided "it would require a regular and steady Course of Observations on a Number of Winters in the different Parts of the Country . . . to obtain full Satisfaction on the Point". By contrast, the American writer Noah Webster needed no further evidence to refute the idea of a permanent warming trend. In his essay "On the Supposed Change in the Temperature of Winter," first read before the Connecticut Academy of Arts and Sciences in 1799, he provided one of the more comprehensive and thoughtful interventions in these debates. Reviewing the succession of authors on the topic, from Antiquity to the turn of the century, Webster "questioned the fact" that warmer winter temperatures in the Northern Hemisphere would be everlasting. He instead attributed any "diminution" in severe weather to mere variability, singling out Williams for having "run into the error . . . of taking the accounts of a few *severe winters* as descriptions of *ordinary winters*".[13]

Even when they did not name their sources, these late-eighteenth-century debates about the reality, causes, or perpetuity of climate change were heavily informed by or formulated in response to Buffon's theoretical synthesis in *Des époques de la nature*, published in 1778 as an instalment of his widely read multivolume *Histoire Naturelle.* In it, Buffon integrated several strands of thinking about climatic instability: (1) local attestations about anthropogenic climate change in European colonies of settlement; (2) historical explanations of the differences between continental climates at the same latitude on either side of the Atlantic; and (3) scientific inferences from mineralogical, fossil, and skeletal remains that offered clues about the strikingly different biogeography of the distant past. Although building on discussions that had been developing over the seventeenth and eighteenth centuries, Buffon's sweeping narrative went further in providing a daringly secular account of the origins of the planet and its natural diversity. Just as with human history, Buffon argued, natural historians had to research "the archives of the world". The Bible reckoned the earth to be 7,000 years old, but in order to explain a range of perplexing natural phenomena, he found it necessary to radically expand the chronology of earth history to between 75,000 and 77,000 years, divided into epochs marking profound ecological revolutions. Buffon proposed that the earth and its various environments took shape over the course of seven epochs. In the beginning, the entire globe had been very hot, in excess of the hottest temperatures in the eighteenth-century tropics. In succeeding epochs the surface became covered with water and, eventually, the climate began to cool at the poles.

As global cooling continued, Buffon argued, discrete bodies of land and water took shape as oceans, continents, and glaciers; humans, creatures, and other life forms began to emerge and disappear, migrate towards the equator, disperse, and differentiate. In the seventh and final epoch, the earth's climate continued to be affected by the process of refrigeration (*refroidissement*) that was especially evident in the uncultivable, uninhabitable polar and alpine regions covered with glaciers. But, paradoxically, the climate in other regions was milder than might be expected, because during this most recent span of millennia, he declared, temperate climates were in significant part an artefact of human ingenuity and labour. Nature had become subject to the improving "power of humankind":

> the entire face of the earth today bears the imprint of the power of man, which, although subordinate to that of nature . . . is a wonderful assistant . . . it is with our hands that [the earth] developed to its full extent, and she came by degrees to the point of perfection and magnificence which we see today.

If it were true that global cooling was the inexorable natural force that largely determined the climate and its history, then only the countervailing force of anthropogenic warming could explain why the entire world was not covered in ice in his lifetime.[14]

Banks was well versed in Buffon's grand view of climate history and learned debates about its merits. He also knew about its influence on a range of corroborating and alternative accounts of climatic amelioration. Samuel Williams and John Mervin Nooth were just two of many naturalists who wrote directly to him in the late eighteenth century to describe and offer scientific explanations of the warming trend that had developed in their corners of North America over the colonial period. Williams wrote to Banks in 1789 to ask if the Royal Society would publish two treatises he had written, one on the nature of heat in the earth's core and another on heat in comets. He also informed Banks about his fieldwork on the relationship between ground, deep well water, and air temperatures measured during his travels throughout the length and breadth of New England – from Long Island Sound to Vermont's border with Quebec and from Boston to the Hudson River. Comparing these measurements, Williams observed

> the very rapid alteration which is taking place in the climate of the country. I have been much engaged in making observations and experiments on this subject. They have afforded me farther information respecting the heat of the earth: and from the alteration which has taken place in the climate of all countries in which philosophical observations have been made, I am convinced that the heat of the earth has been gradually increasing.

In his answer, Banks offered no response to the comments about climate warming. He merely acknowledged receipt of Williams's treatises but cryptically explained that they would not be published in the *Philosophical Transactions* because they had already been superseded by "certain discoveries not yet come to your side of the water" (it is unknown to which discoveries Banks was referring).[15]

Just a few years later, Banks archived a similar paper written by Sir John Leslie, a professor of mathematics and natural philosophy at Edinburgh. Leslie studied caloric theory and the physics of light; he later invented several instruments related to this work, including a differential hygrometer and photometer. His paper "On Heat and Climate" attempted to apply his laboratory research on heat to understanding climate and weather because, as he stated, "The more

I reflect on the proposition, that climates are growing gradually warmer, the more am I convinced of its reality". Leslie dismissed the impact of anthropogenic forces – "the clearing of the forests, the draining of the marshes – on the "general amelioration of the globe", although he allowed that "America forms a curious exception to the general rule". The paper was read aloud at a meeting of the Royal Society in February or March 1793, but it was not published in the *Philosophical Transactions*.[16]

Until the turn of the century, all we know about Banks's assessment of scientific accounts or evidence of climate change is that he did not privately or publicly record it. Perhaps he was too busy promoting a multitude of competing concerns in which he was involved, ranging from the African Association to the Royal Mint. We can exclude the possibility that he had religious objections to secular accounts of climate change as a form of heresy. Since Banks was never particularly outspoken about his Christian faith, his reticence to discuss climate change was unlikely to have been a reflection of pious conservatism. Scholars have surmised that he was probably a Deist, a set of beliefs that would not have interfered with conjectures about a geological past deeper than the biblical account allowed. When he learned that church ministers in Edinburgh had censured and attempted to block Leslie's appointment to the university for his support of Hume's writings, for example, Banks reassured Leslie: "Surely a Man may fulfill his duty to his Creator, & render his Redeemer propitious without assenting unconditionally to every undigested Tenet which our half-informed Predecessors have left behind as a legacy to their more enlightened successors".[17]

It was the persistence of uncommonly cool English weather that sufficiently excited Banks's interest in the implications of climatic instability. In the 1790s, he became troubled by what he experienced as notably colder, shorter summers, conditions that negatively affected the domestic agricultural economy, including the rents collected from tenants farming his Lincolnshire properties. The disconcerting prospect of continued harvest shortfalls of wheat and other traditional staple grains became more alarming with the onset of the Napoleonic Wars. As a result, members of the Board of Agriculture and Royal Horticultural Society encouraged their members to experiment with breeding more cold-tolerant crops. Banks's own efforts toward this goal focused on frost-resistant wild rice imported from Canada to his suburban London estate, where he hoped to acclimatize the plant to "the ungenial springs, the chilly summers, and the rigorous winters, by which, especially for some years past, we have been perpetually visited". His initial success with this project seemed to "point out the road" to one way in which English farmers might adapt to protracted shifts in local weather.[18]

In December 1805, Banks described this experiment to an audience assembled at the Royal Horticultural Society. This essay, entitled "Some Hints Respecting the Proper Mode of Inuring Tender Plants to Our Climate", is the earliest indication that he was thinking about climate as an unstable phenomenon. In it he also made an allusion to a theory of species migration due to climate change, explaining that he had "some reason to believe, that every" plant acclimatized to English weather "was originally the native of a cold climate, though introduced to us through the medium of a warm one". These passing remarks probably derived from Buffon's theory of global cooling, but Banks might have also had in mind John Leslie's lesser-known revisionist theory of glacial advance and retreat detailed in *An Experimental Inquiry into the Nature and Propagation of Heat*, which had been published the previous year and which Banks had read. Like his earlier paper "On Heat and Climate", Leslie claimed that this work's main objective "was to understand the nature of radiant heat", but he was also interested in applying these theoretical insights to reconstructing the history and predicting the future of the earth and its atmospheres. Ultimately, his ambition was to establish a new scientific discipline: research on the principles of heat and light would "prepare a solid foundation for erecting a system of Meteorology". The "circumstances relative to climate", he emphasized, "deserve particular investigation".[19]

In an exceptionally long footnote, Leslie laid out a deep history of the earth. Echoing *Des époques de la nature*, he declared,

> [T]he history of our species with all its busy and tragic scenes shrinks into a point in comparison of those vast cycles which are familiar to the mind of the geologist or astronomer. Man, the last and most perfect of Nature's works, is only a recent inhabitant of this globe.

He continued to believe it was "unquestionable that the climate over the whole of Europe has assumed a milder character". Despite the similarity of his and Buffon's timescales and their agreement on recent global warming, Leslie argued that glacial retreat was the predominant cause of contemporary climate change. While it was certain that "the temperature of our globe increases", humanity's efforts to improve land had made only a minor contribution to this "extremely slow" process, which began long before the development of settled agriculture. "No mighty change has actually taken place within the period of authentic history", Leslie wrote. His methodology for arriving at this conclusion was to combine theory developed from laboratory experiments with his observations about the cause for "the contraction of glaciers" that he witnessed first-hand during a hiking trip in the Swiss Alps in 1796. "The alternate process of thawing and freezing" that occurred every spring and winter there caused avalanches of compacted snow that accumulated into "icy zones" on mountainsides and valleys at the highest elevations. But true glaciers, he explained, "were formed only from avalanches of a rarer and more formidable kind", that is, when an icy zone grew too heavy and disintegrated into massive shards. When these "vast irregular masses of ice" tumbled to lower, warmer elevations, glaciers were born and, at the very same time, began to "suffer a continual diminution". Several centuries "may be required to melt the whole", but the unmistakable marks of glacial retreat coupled with the advance of vegetation could be seen across the Alps. Leslie correlated this visual evidence to his impression that "every object around begins already to assume the character of an Italian climate". The Arctic was the "only apparent exception" to this widespread pattern, he admitted, at least insofar as polar conditions could be extrapolated from the "inclement skies and penurious soil" found in Iceland and Greenland.[20]

When Banks originally read *An Experimental Inquiry*, he was unimpressed with Leslie's "book on the subject of heat & c., which certainly contains many interesting experiments & much bad reasoning". Nevertheless, a little over a decade later, such accounts gained new relevance for him. Four days after he received Scoresby's description of melting polar ice, he wrote to a colleague in the horticultural society claiming that he "had always attributed the increasing Coldness of our Climate to the increase of Polar Ice". But, he continued, this certitude seemed to be overturned by recent reports of conditions in the Alps and Arctic:

> The newspapers during the whole summer & autumn have been full of inundations in all Countries where Rivers are fed from Icy mountains, & have told us that the Coast of West Greenland, which has not been free from ice for 150 years, has been this year seen & found clear of all obstructions for many leagues. The Atlantic has been unusually clogged with Islands of Ice. Possibly I am too sanguine, but . . . I feel a hope that we shall be indulged with better Springs than have lately been provided for us.

What exactly did he mean, and from what theory did he derive his speculation that crumbling glaciers suggested warmer springs to come? Banks was of course aware of other relevant currents in geology, including through his honorary membership in the London-based Geological Society and the Edinburgh-based Wernerian Society, founded in 1807 and 1808, respectively. While these groups were interested in Arctic geography, they increasingly favoured field-based and descriptive empirical work over grand geo-historical theorizing. Their members mainly focused on rock layers, minerals, and fossils gathered in clearly delineated locations or studied specific mountains, volcanic eruptions, and earthquakes. Relatively few scientists in this period focused on the history of glacial formation and change, let alone its implications for what came to be known much later as the oceanic or the atmospheric sciences. Though neither Buffon's nor Leslie's theories fully explained what Scoresby had seen in the Arctic – since both assumed that polar climates were irreversibly frigid – in 1817 they were still among the only ones upon which Banks could have drawn that dealt with glacial history.[21]

Whatever scientific accounts of glacial retreat Banks was weighing in his mind, he maintained that the phenomenon remained "inexplicable at present to us". Despite this uncertainty, he began to suspect that England's cooler climate was only a side effect of evaporating glaciers. Soon he began to broadcast to administrators and politicians that he had gathered "ample proof that new sources of warmth have been opened and give us leave to hope that the Arctic seas may at this time be more accessible than they have been for centuries past". At the end of November 1817 he wrote to the First Lord of the Admiralty, explicitly asserting "that a considerable change of climate . . . must have taken place in the circumpolar regions by which the severity of the cold, that has for centuries past enclosed the seas in the high northern latitudes in an impenetrable barrier of ice, has been, during the last two years, gradually abated". In lobbying the Admiralty, Banks had argued that among the greatest dividends of Arctic expeditions would be "the extension of useful knowledge and the general benefit of mankind". But, in the meantime, taking advantage of climate warming trumped finding a cogent scientific explanation for it. Within days, British newspapers announced, "[O]wing to the statements of the Greenland Captains respecting the diminution of the polar ice, the Royal Society had applied to Ministers to send out vessels for discovery in the Polar Seas".[22]

Scoresby assumed he would command the first expedition, not only because he had delivered the news about an open Arctic but also because Banks had led him to believe that his combination of maritime experience and scientific credentials made him the perfect candidate. Since his days as a student at Edinburgh – where he had attended lectures given by John Leslie, among others – Scoresby had gained recognition in British scientific circles for his research in Arctic natural history. In 1807, his father introduced him to Banks, "a personal intercourse", Scoresby later recalled, "from which I derived very great mental advantages". On shipboard during whaling seasons, he kept meteorological journals that made use of Luke Howard's terms for classifying clouds; noted the quality of blubber in different whale species; created some of the earliest taxonomical drawings of snowflakes (or "snow crystals"); and experimented with various methods for measuring deep seawater temperatures, hoping ultimately to be able to ascertain the mean temperature of the Arctic Ocean. Besides publishing short essays on these topics in the Wernerian Society's journal, he was also at work on a book about the Far North. In November 1817 Banks wrote to him, "I hope you proceed quickly with your intended work on the Polar Seas. I am impatient to see it after having so much profited by your Essay". Scoresby therefore trusted Banks to support his intellectual pursuits and honour his wish to have "the command of an expedition for discovery, instead of fishing" by empowering him to locate "a northern passage to India" (even if, as he warned Banks, "once accomplished it might not again be practicable in ten or even twenty years").[23]

Despite Banks's earlier assurances, by early December 1817 the Navy had already finalized its appointments: they bypassed Scoresby in favour of the less scientifically inclined officers David Buchan and John Ross. Banks apologized to a dismayed Scoresby that, technically, the Navy "could not employ any but their own officers as leading men". In fact, Banks had allied himself with his friend John Barrow, the careerist second secretary of the Admiralty and a Northwest Passage enthusiast. Barrow was certain that the North Pole was entirely navigable and knew that Scoresby disagreed. "Though the Polar seas were navigable in an uncommon degree last summer", Scoresby counselled Banks, "I conceive it very uncertain whether the ice may yet remain the same, and whether the navigation of these seas still continues equally open. As to reaching the Pole, I confess myself sceptical". In a less flattering formulation, he belittled the idea of reaching the actual North Pole as "the frenzied speculation of a disordered fancy". Insulted by Scoresby's views and finding his meticulous empiricism to be a nuisance, Barrow made sure to exclude him from consideration.[24]

Meanwhile, Banks continued to seek out Scoresby's opinion about northern climate change. As Ross and Buchan's ships were being outfitted in February 1818, Banks asked what he knew about sightings of exceptional "ice islands" near the Newfoundland coasts, which experienced local fishermen claimed were the largest they had seen in fifty years. If this information could be confirmed, Banks would finally be persuaded that glacial retreat explained the odd English weather at the turn of the century. Abandoning his circumspection on the question of climate change, Banks wrote, "Should this be the case, it will account fully for the diminished heat of our summers".[25] When Scoresby returned from whaling at the end of summer 1818, Banks despaired that the Arctic ice had returned to normal levels. Reiterating Scoresby's news, he wrote to him with disappointment:

> By your account it should seem that the coast of Greenland on the E side is this year as inaccessible as it has been for the last 400 years and that the absence of the ice last year has been only an accidental variation of the climate or rather of the season and I fear that it does not prove as we hoped it did an indication of a diminution of the ice in the Polar basin.[26]

Banks's expectations to satisfy his speculations about climate change were dashed, but only briefly. Scoresby immediately responded to clarify that conditions were much the same as they had been in 1817:

> I perceive I have been misunderstood as to the state of the polar ice during the summer of the present year. When I remarked that the coast of Greenland was inaccessible from latitudes 74 to 80, at least, after the middle of June, I did not mean to imply that it was owing to any increase of ice between these parallels since last year, but that it was occasioned by the packing of that ice together . . . The quantity of ice on the east side of Greenland during the present year I take as very nearly the same as last. If there was any increase it must have been very unimportant.[27]

Though he remained uncertain about the future conditions of the ice and could not guarantee success, Scoresby had been persuaded that "a north-west passage exists" (or, at least, he had become willing to adopt Barrow's views so as to secure a government commission to support his Arctic research). Scoresby advised Banks that the Navy should sponsor "repeated trials" of the search for a maritime route to the Pacific, treating it as a worthy scientific experiment on a grand scale. In addition, he proposed that his own contributions toward that objective could be to produce a series of maps charting the maximum and minimum extent of Arctic ice.[28]

Scoresby came to regret that hasty reply. A string of frustrated missions in the following decade vindicated his suspicion that the polar thaw was temporary.

Yet to his chagrin, in the 1830s John Ross publicly scapegoated Scoresby for misguiding the Navy, compelling him to defend his limited role as a mere informant to Banks. In a book about the Arctic published in 1831, John Leslie exculpated his former student, explaining that Scoresby accurately described the sea ice as he witnessed it but never suggested that the melting would be constant. By contrast, Leslie implicitly rebuked Banks (who died in 1820) and Barrow (to whom, inexplicably, the book was dedicated) as "some persons" who "imagined that the mountains or islands of ice, which are occasionally drifted into the Atlantic Ocean, must be sufficient, by their frigorific influence, to modify the character of our climate".[29] While Leslie continued to believe in a very gradual, subtle tempering of the European climate due to the "periodical formation and destruction of the Polar ice", he also reaffirmed that the Arctic Circle was itself impassable: "There is nothing certainly in their history which betrays any radical or permanent change in the climate of the Arctic regions. The same continent of ice still remains during the far greater part of the year, to bar the access of the navigator to the Pole".[30]

Long since then, modern climate scientists have developed a more precise explanation for the unique collision of geophysical events affecting the Arctic in the 1810s. As Noah Webster and Leslie had argued, perceptible changes in weather patterns had been signals of periodical variability and unique but fleeting circumstances rather than permanent climate change. Although polar sea ice is millions of years old, its surface layers are so sensitive to climate change that they have always expanded and contracted, such as occurred during the Little Ice Age that encompassed the entire early modern period through the nineteenth century. Expanding Arctic glaciers and sea ice caused unusual or severe weather around the world; across the Northern Hemisphere, there were subtle but significant drops in prevailing seasonal temperatures, especially in coastal regions, which were subject to prolonged hard winters and abbreviated, cool, and wet summers. Despite that Little Ice Age conditions were waning in 1816, it proved to be a year of exceptionally harsh weather. The harvest killing frosts of this so-called year without a summer were one of the complex global effects of a distant volcanic eruption of Mount Tambora in 1815, which reverberated through the rest of the decade. In the North Atlantic Ocean, intensified colder conditions across the temperate latitudes coincided with the unusually warm temperatures in the Arctic that experienced mariners like Scoresby regarded with such puzzlement.[31]

If current scientific knowledge of historical climate fluctuations clarifies such anomalies, it does not help us interpret the politics of debates about climate science in the past or judge their significance for current debates. For anyone looking to history for examples of environmental governance, eighteenth- and early-nineteenth-century Britain offers a uniquely informative model because of the global reach of its empire and extensive involvement of scientists in it. In this period, scientists alert to the dynamic, unpredictable nature of climate were beginning to develop explanations for it. They were seriously engaged in delineating the history of the earth, understanding prevailing weather patterns, and predicting future climatic conditions. They were hardly naive about the stakes involved in publicizing their theoretical work, the ways in which their scientific endeavours were shaped by the political economy of patronage, or the priority for government to pursue strategic objectives over rigorous research. Officials' desires to act on the semblance of climatic change proved more important than naturalists' best efforts to understand its causes and limits. Through the remainder of the century, Britain continued to send explorers into the Arctic to chase after the Northwest Passage even in the face of mounting evidence that it was impossible. In the twenty-first century, this is no longer the case. The rate of melting glaciers and sea ice, largely as a result of anthropogenic forces, is more rapid than at any other time in the last several thousand years. In the wake of glacial retreat, European and North American nations are scrambling once again to colonize the region. As an ice-free Arctic becomes a genuine probability, it is worth remembering the uses and misuses of climate science in the past.[32]

Notes

* Thanks very much to John Gascoigne for sharing his transcript copies of letters from the Scoresby Archive and to Neil Chambers for comments on a draft of the essay.

1 William Scoresby, Jr. (1789–1857), not to be confused with his father William Scoresby, who was also a whaling captain. Tom Stamp and Cordelia Stamp, *William Scoresby, Arctic Scientist* (Whitby: Caedmon of Whitby Press, 1976), 66 (perfectly void; mystery); Joseph Banks to William Scoresby, 26 Oct. 1817, Scoresby Archive, Whitby Literary and Philosophical Society, Whitby Museum, Yorkshire, England; Fergus Fleming, *Barrow's Boys* (London: Granta, 1999), 32–33 (attentive observation). Banks wrote that Scoresby's account was first published in a Liverpool newspaper, but I have been unable to locate it. To Banks, he reported '2000 leagues' of melted ice – approximately 6000 square miles.

2 Banks to Scoresby, 22 Sept. 1817, in *The Letters of Sir Joseph Banks: A Selection, 1768–1820*, ed. Neil Chambers (London: Imperial College Press, 2000), 329 (matter); George-Louis Leclerc, comte de Buffon, *Histoire Naturelle: générale et particulière. Des époques de la nature*, vol. 5 (Paris, 1778), 222 (solid ice); Anon, *The Journal of a Voyage Undertaken by Order of His Present Majesty, for Making Discoveries Towards the North Pole, by the Hon. Commodore Phipps* (London, 1774), 118 (certainly exists); William Scoresby, Jr., 'On the Greenland or Polar Ice (Read 11 March 1815)', *Memoirs of the Wernerian Natural History Society*, vol. 2 (Edinburgh, 1818), 261–338, 338 (improbable); Banks to Scoresby, 17 Nov. 1817, Scoresby Archive (maritime science). For contemporary accounts of the Northwest Passage expeditions, see John Barrow, *A Chronological History of Voyages into the Arctic Regions; Undertaken Chiefly for the Purpose of Discovering a North-East, North-West, or Polar Passage between the Atlantic and Pacific* (London, 1818); John Ross, *A Voyage of Discovery . . . for the Purpose of Exploring Baffin's Bay and Inquiring into the Probability of a Northwest Passage* (London, 1819); W. Edward Parry, *Journal of a Voyage for the Discovery of a Northwest Passage from the Atlantic to the Pacific Performed in the Years 1819–20* (London, 1821). On science at sea in this period, see Simon Naylor, 'Log Books and the Law of Storms: Maritime Meteorology and the British Admiralty in the Nineteenth Century', *Isis* 106, no. 4 (2015): 771–797. Locating a Northwest Passage was one of Captain Cook's directives for his third circumnavigation in the late 1770s, after which government-sponsored initiatives of this sort focused on overland and riverine routes through, as opposed to maritime routes around, northern North America until the episode recounted in this essay. See Glyndwr Williams,

Voyages of Delusion: The Search for the Northwest Passage in the Age of Reason (New Haven: Yale University Press, 2003); and *Arctic Ambitions: Captain Cook and the Northwest Passage*, ed. James K. Barnett and David Nicandri (Anchorage: Anchorage Museum, 2015).

3 John Gascoigne, *Science in the Service of Empire: Joseph Banks, the British State, and the Uses of Science in the Age of Revolution* (Cambridge: Cambridge University Press, 1998); John Gascoigne, *Joseph Banks and the English Enlightenment: Useful Knowledge and Public Culture* (Cambridge: Cambridge University Press, 2003); Richard Drayton, *Nature's Government: Science, Imperial Britain, and the 'Improvement' of the World* (New Haven: Yale University Press, 2000).

4 For the cultural history of Arctic science in the nineteenth and twentieth centuries, see Trevor H. Levere, *Science and the Canadian Arctic: A Century of Exploration, 1818–1918* (Cambridge: Cambridge University Press, 1993); Mark Carey, 'The History of Ice: How Glaciers Became an Endangered Species', *Environmental History* 12, no. 3 (July, 2007): 497–527; *Narrating the Arctic: A Cultural History of Nordic Scientific Practices*, ed. Michael Bravo and Sverker Sörlin (Canton: Watson Publishing International, 2002); and Theodore Binnema, *Enlightened Zeal: The Hudson's Bay Company and Scientific Networks, 1670–1870* (Toronto: University of Toronto Press, 2014). For the debate about the novelty of these concerns in the Anthropocene, see Dipesh Chakrabarty, 'The Climate of History: Four Theses', *Critical Inquiry* 35 (Winter, 2009): 197–222; Dipesh Chakrabarty, 'Climate and Capital: On Conjoined Histories', *Critical Inquiry* 41, no. 1 (Autumn, 2014): 1–23; and Christopher Bonneuil and Jean-Baptiste Fressoz, *The Shock of the Anthropocene: The Earth, History, and Us* (New York: Verso, 2016).

5 Anya Zilberstein, *A Temperate Empire: Making Climate Change in Early America* (Oxford: Oxford University Press, 2016).

6 For the sake of analysis (and lack of better terminology), we might distinguish between glaciological climate theory derived from the earth sciences as opposed to the much more prevalent and traditional anthropogenic theory based on observations and measurements of environmental changes in cultivated landscapes. It might be worth extending to the climate maritime historians' argument that ocean environments were neither a void nor a resource commons but instead extraterritorial zones subject to commercial exploitation, government intervention, and international competition. See Lauren Benton, *A Search for Sovereignty: Law and Geography in European Empires, 1450–1900* (Cambridge: Cambridge University Press, 2010); W. Jeffrey Bolster, 'Opportunities for Marine Environmental History', *Environmental History* 11, no. 3 (July, 2006): 567–597; Helen Rozwadowski and Michael S. Reidy, 'Spaces in between: Science, Ocean, Empire', *Isis* 105, no. 2 (June, 2014): 335–337.

7 Clarence J. Glacken, *Traces on the Rhodian Shore: Nature and Culture in Western Thought From Ancient Times to the Present* (Berkeley: University of California Press, 1967); Karen Ordahl Kupperman, 'Climate and Mastery of the Wilderness in Seventeenth-Century New England', in *Seventeenth-Century New England*, ed. David D. Hall and David Grayson Allen (Boston: Colonial Society of Massachusetts, 1984), 3–38; Richard H. Grove, *Green Imperialism: Colonial Expansion, Tropical Island Edens and the Origins of Environmentalism* (Cambridge: Cambridge University Press, 1995), Chapters 4 and 5; James R. Fleming, *Historical Perspectives on Climate Change* (Oxford: Oxford University Press, 1998); Zilberstein, *A Temperate Empire.*

8 James Franklin, *The Philosophical and Political History of the Thirteen United States of America* (London, 1784), 43–44.

9 'Change of Climate in North America and Europe', n.d. Box 1, Folder 10, Samuel Williams Papers, Harvard University Archives; 'Natural History', Folder 65, Samuel Williams Family Papers, Ms AM 2624, Houghton Library, Harvard University; Samuel Williams, *The Natural and Civil History of Vermont* (Walpole, NH, 1794), 57 (new country); House of Commons, *Report from the Committee on the Petition of the Court of Directors of the Sierra Leone Company* (London, 1802), 744; Jan Golinski, 'American Climate and the

Civilization of Nature', in *Science and Empire in the Atlantic World*, ed. Nicholas Dew and James Delbourgo (New York: Routledge, 2008), 153–174; Richard W. Judd, *The Untilled Garden: Natural History and the Spirit of Conservation in America, 1740–1840* (Cambridge: Cambridge University Press, 2009), 228–235; Fleming, *Historical Perspectives on Climate Change*, 11–32.

10 *Peter Kalm's Travels in North America*, ed. and trans. Adolph B. Benson (New York: Wilson-Erickson, 1964); Fredrik A. Jonsson, 'Climate Change and the Retreat of the Atlantic: The Cameralist Context of Pehr Kalm's Voyage to North America, 1748–51', *The William and Mary Quarterly* 72, no. 1 (January, 2015): 99–126.

11 John Mitchell, *Present State of Great Britain and North America*, 2 vols., vol. 1 (London, 1767), 167; Edward Long, *A Free and Candid Review of a Tract, Entitled Observations on the Commerce of the American States* (London, 1784), 15–16, 21, 44 (emphasis in the original). On Priestley, phlogiston, and aerial reform see Jan Golinski, *British Weather and the Climate of Enlightenment* (Chicago: University of Chicago Press, 2007), 159–166.

12 Fredrik A. Jonsson, *Enlightenment's Frontier: The Scottish Highlands and the Origins of Environmentalism* (New Haven: Yale University Press, 2013), Chapter 3; Gillen D'Arcy Wood, *Tambora: The Eruption That Changed the World* (Princeton, NJ: Princeton University Press, 2014); Nooth to Banks, 2 Jan. 1792, 'Recueil de lettres autographes signées du docteur J. Mervin Nooth, 1789–1799, 1902', 038–03–02–08, Section des archives, Ville de Montréal, Archives de Montréal, Québec (fictive Quebec volcano).

13 Benjamin Franklin to Ezra Stiles, 29 May 1763, in *The Papers of Benjamin Franklin, vol. 10, January 1, 1762, through December 31, 1763*, ed. Leonard W. Labaree (New Haven: Yale University Press, 1959), 264–267; Noah Webster, 'A Dissertation on the Supposed Change in the Temperature of Winter, (Read before the Connecticut Academy of Arts and Sciences, 1799)', *Memoirs of the Connecticut Academy of Arts and Sciences*, vol. 1 (New Haven, 1810), 1–68, 13, 18 (emphasis in the original).

14 Buffon, *Des époques*, 1 ("*dans l'histoire naturelle, il faut fouiller les archives du monde*"), 67 and 242 (age of earth), 225 ("*Lorsque la puissance de l'homme a secondé celle de la nature*"), 237 ("*enfin la face entière de la terre porte aujourd'hui l'empreinte de la puissance de l'homme, laquelle, quoique subordonnée à celle de la nature, souvent a fait plus qu'elle, ou du moins l'a si merveilleusement secondée, que c'est à l'aide de nos mains qu'elle s'est développée dans toute son étendue, et qu'elle est arrivée par degrés au point de perfection et de magnificence où nous la voyons aujourd'hui*"); Martin J.S. Rudwick, *Bursting the Limits of Time: The Reconstruction of Geohistory in the Age of Revolution* (Chicago: University of Chicago Press, 2005), 75–80.

15 16 September 1789, Williams to Banks, Add MS 8097.358, British Library, London; 16 September 1789, Williams to Banks (Williams' copy), 2–28, Samuel Williams Papers, Special Collections, University of Vermont Library, Burlington; 20 July 1790, Banks to Williams (Microfilm copy), Series 73.032, Section 15, Papers of Sir Joseph Banks, State Library, New South Wales.

16 John Leslie, 'On Heat and Climate', *Annals of Philosophy* 14 (1819): 5–27: 21, 23. Leslie finally published the paper in 1819 in a Scottish scientific journal. For a consideration of the theoretical aspects of Leslie's work, see J.L. Heilbron, 'Weighing Imponderables and Other Quantitative Science around 1800', *Historical Studies in the Physical and Biological Sciences* 24, no. 1 (1993): 1–277, 117–127; and Richard G. Olson, 'Count Rumford, Sir John Leslie, and the Study of the Nature and Propagation of Heat at the Beginning of the Nineteenth Century', *Annals of Science* 26, no. 4 (1970): 273–304.

17 19 April 1805, Banks to Leslie, in Chambers, *Letters of Sir Joseph Banks*, 264; Gascoigne, *Banks and the English Enlightenment*, 41–55; Patrick O'Brian, *Joseph Banks: A Life* (Chicago: University of Chicago Press, 1987), 106–107. For a broader discussion of the relationship between theology and Enlightenment climate theory, see Lydia Barnett, 'The Theology of Climate Change: Sin as Agency in the Enlightenment's Anthropocene', *Environmental History* 20 (2015): 217–237.

18 Joseph Banks, 'Some Hints Respecting the Proper Mode of Inuring Tender Plants to Our Climate', *Transactions of the Horticultural Society*, 1 (3rd ed., London, 1820), 21–25; Anya Zilberstein, 'Inured to Empire: Wild Rice and Climate Change', *The William and Mary Quarterly* 72, no. 1 (January, 2015): 125–156.

19 Banks, 'Some Hints', 21; John Leslie, *An Experimental Inquiry into the Nature and Propagation of Heat* (London, 1804), x–xiii; [n.d.] April 1804, Banks to Count Rumford, in Chambers, *Letters of Sir Joseph Banks*, 256–257.

20 Leslie, *An Experimental Inquiry*, 181–182, 536–537, 539–541. Leslie also believed his account of how glaciers formed explained why, according to him, there were none in the Andes.

21 [n.d.] April 1804, Banks to Count Rumford, in Chambers, *Letters of Sir Joseph Banks*, 256–257 (bad reasoning); 26 September 1817, Banks to Thomas Andrew Knight, in Chambers, *Letters of Sir Joseph Banks*, 331. On the Geological Society, see Simon J. Knell, 'The Road to Smith: How the Geological Society Came to Possess English Geology', *The Geological Society* (Special Issue: The Making of the Geological Society of London) 317 (2009): 1–47; Martin J.S. Rudwick, 'The Early Geological Society in Its International Context', *The Geological Society* (Special Issue: The Making of the Geological Society of London) 317 (2009): 145–153; and C.D. Yonge, *The History of the British Navy from the Earliest Period to the Present Time*, 2 vols. (London: Richard Bentley, 1863), 344–347. Rumford and Leslie had a public argument over whether Leslie's *An Experiment* was in fact plagiarized from Rumford's laboratory work in the physics of heat and light. It should be noted that, although Rumford was interested in the nature of solar rays on temperature differences in different climate zones, he wrote nothing about climate history. See Comte de Rumford, *Memoires sur la chaleur* (Paris, 1804).

22 16 December 1818, Duke of Somerset to Banks, Correspondence Concerning Iceland, Written to Sir Joseph Banks, Special Collections, University of Wisconsin Memorial Library, Madison, http://digital.library.wisc.edu/1711.dl/HistSciTech.BanksJ (Somerset was a member on the Admiralty's Board of Commissioners from 1816–1818 and an MP for Monmouth); 20 November 1817, Banks to Richard Saunders Dundas, in Chambers, *Letters of Sir Joseph Banks*, 334–335 (considerable change, ample proof); William Scoresby, Jr., 'On Some Circumstances Connected with the Original Suggestion of the Modern Arctic Expeditions: Communicated by Rev. W. Scoresby', *The New Edinburgh Philosophical Journal* 20 (1836): 93–100, 95 (newspapers).

23 Stamp and Stamp, *William Scoresby*, 33 (intercourse); William Scoresby, Jr., 'Meteorological Journals Kept . . . Greenland 1807, 1808, 1809 (Read 3 Feb. 1810)', *Memoirs of the Wernerian Natural History Society* 1 (1811): 249–257; William Scoresby, Jr., 'Account of the *Balaena Mysticetus*, or Great Northern or Greenland Whale (Read 3 Feb. 1810)', *Memoirs of the Wernerian Natural History Society* 1 (1811): 578–586; William Scoresby, Jr., 'Meteorological Observations on a Greenland Voyage in the Ship Resolution, in the Year 1810 (Read 24 Nov. 1810)', *Memoirs of the Wernerian Natural History Society* 1 (1811): 609–615; 8 September 1810, Banks to Scoresby, and 17 November 1817, Banks to Scoresby, Scoresby Archive, Whitby; Fleming, *Barrows' Boys*, 32–33 (25 November 1817, Scoresby to Banks letter quoted in full); Michael Bravo, 'Geographies of Exploration and Improvement: William Scoresby and Arctic Whaling, 1782–1822', *Journal of Historical Geography* 32 (2006): 512–538.

24 Stamp and Stamp, *William Scoresby*, 67–68; William Scoresby, Jr., 'On the Approximation Towards the Poles, and On the Possibility of Reaching the North Pole', in *A Voyage to Spitzbergen*, ed. John Laing (2nd ed., Edinburgh, 1818), 158 (fancy). For the political dividends of Barrow and Banks's longstanding alliance, see Gascoigne, *Science in the Service of Empire*, 126–127.

25 10 February 1818, Banks to Scoresby.

26 15 September 1818, Banks to Scoresby.

27 26 September 1818, Scoresby to Banks.

28 10 February 1818, Banks to Scoresby and 15 September 1818, Banks to Scoresby, Scoresby Archive, Whitby (400 years); 26 September 1818, Scoresby to Banks, Correspondence Concerning Iceland, Madison; Stamp and Stamp, *William Scoresby*, 70.

29 John Leslie, 'The Climate of the Polar Regions'.

30 Scoresby, 'On Some Circumstances', 94; John Leslie, 'The Climate of the Polar Regions', in *Narrative of Discovery and Adventure in the Polar Seas and Regions*, ed. John Leslie, Robert Jameson and Hugh Murray (New York, 1831), 44 (defence of Scoresby), 52 (radical).

31 Jan Zalasiewicz and Mark Williams, *Ocean Worlds: The Story of Seas on Earth and Other Planets* (Oxford: Oxford University Press, 2014), 178–180; William R. Barron, '1816 in Perspective: The View from the Northeastern United States', in *The Year Without a Summer? World Climate in 1816*, ed. C.R. Harrington (Ottawa: Canadian Museum of Nature, 1992), 124–144; Wood, *Tambora*, 138–141; Hubert H. Lamb, *The Changing Climate: Selected Papers* (London: Methuen, 1966); Shaun A. Marcott, Jeremy D. Shakun, Peter U. Clark, and Alan C. Mix, 'A Reconstruction of Regional and Global Temperature for the Past 11,300 Years', *Science* 339, no. 1198 (2013), 1198–1201, DOI: 10.1126/science.1228026; John F. Richards, *The Unending Frontier: An Environmental History of the Early Modern World* (Berkeley: University of California Press, 2003), Chapter 1; Paul N. Edwards, *A Vast Machine: Computer Models, Climate Data, and the Politics of Global Warming* (Cambridge: MIT Press, 2010); Thomas M. Wickman, 'Snowshoe Country: Indians, Colonists, and Winter Spaces of Power in the Northeast, 1620–1727' (unpublished PhD dissertation, Harvard University, Cambridge, MA, 2012); Geoffrey Parker, *Global Crisis: War, Climate Change, and Catastrophe in the Seventeenth Century* (New Haven: Yale University Press, 2013); Sam White, 'The Real Little Ice Age', *Journal of Interdisciplinary History* 44 (2013): 327–352.

32 On Russian and British imperial competition over the North Pacific, see James Delgado, 'The Continuing Quest: The Lure of the Northwest Passage in the Nineteenth and Twentieth Centuries', in James K. Barnett and David Nicandri, *Arctic Ambitions: Captain Cook and the Northwest Passage* (Anchorage: Anchorage Museum, 2015), 313–329; and Ryan T. Jones, *Empire of Extinction: Russians and the North Pacific's Strange Beasts of the Sea, 1741–1867* (Oxford: Oxford University Press, 2014); Zalasiewicz and Williams, *Ocean Worlds*, 179; Natalia Vizcarra, 'February Continues Streak of Record Low Arctic Sea Ice Extent', March 10, 2016, *NASA: National Snow and Ice Data Center*. http://climate.nasa.gov/news/2411/.

8 'Vast factories of febrile poison'

Wetlands, drainage, and the fate of American climates, 1750–1850

Anthony E. Carlson

Environmental historians have largely ignored the 1790s American climate crisis. In 1793, yellow fever, which had been absent in the former British mainland colonies for three decades, erupted across several cities, including New York and Philadelphia, the young republic's capital. Philadelphians suffered the worst. By the time the epidemic faded, following the first frost, approximately 5,000 Philadelphians – roughly 10 per cent of the city's population – lay dead. During the next dozen years, yellow fever exploded up and down the Atlantic seaboard, killing thousands in cities ranging from as far north as Portsmouth, New Hampshire, to as far south as Savannah, Georgia. The disease also struck New Orleans in 1796, 1799, 1800, and 1804. The cycles of outbreaks triggered a lengthy and contentious medical debate about the disease's foreign or domestic etiological origins, but they also challenged the 'widespread consensus' that enlightened European methods of land management, including intensive monoculture, deforestation, and wetlands drainage, had moderated the sickly American climates, which Americans conceptualized as infinitely malleable microenvironments where the interaction of air, water, and soil influenced the health of living organisms.[1]

Yellow fever's resurgence shattered the Enlightenment-inspired belief that European strategies of land management could improve unhealthy and harsh climates, making them fit for European bodies. Drawing heavily on the resurgent interest in Hippocratic medical theories, the growth of medical meteorology, and pneumatic chemistry experimentation, the Enlightenment conviction that civilized societies should reduce nature to a state of passivity by improving unhealthy climates through drainage and soil cultivation constituted the fountainhead of early American environmental thought. Focused on concepts such as wilderness, romanticism, ecological degradation, industrialization, and incipient conservationism, environmental historians have downplayed this rich and pervasive discourse, largely ignoring the continuity of many premodern climactic theories throughout the nineteenth century.

This chapter argues that American physicians, political leaders, naturalists, and rural press editors responded to yellow fever's return to the United States by identifying the distribution of water on the earth's surface as the primary barometer of healthy climates. Arguing that many diseases occupied particular geographic spaces, they insisted that the decomposition of animal and vegetable matter in marshes and other landscapes characterized by an abundance of surface water produced miasmas, 'miasmata', and other unwholesome airs that, once inhaled, hastened fevers and sometimes even death. Reframing pathogenic climates from the perspective of marshes and wetlands, they insisted that it was foolish, sometimes even fatal, to leave the environment in its original, unimproved condition. As a vector of civilization and uncorrupted airs, drainage became the panacea for promoting pure airs and healthy bodies. Ultimately, the collective response to the 1790s climate crisis forged a common set of assumptions about responsible land management that elevated drainage into a paramount public policy objective, shaping the course of nineteenth-century water resources management and early American environmental thought.

A violent and fatal climate

During the earliest stages of European colonization, the North American climate proved perplexing. Unlike Europe and even tropical destinations, North America's weather phenomena were extreme, unpredictable, variable, and violent. In addition to oppressive humidity, extreme temperature fluctuations, intense thunderstorms, hurricanes, waterspouts, and tornadoes, sixteenth- and seventeenth-century colonists found climate's latitudinal definition to be wanting. As the cultivation of tropical crops in the Chesapeake failed and colonists discovered that Nova Scotia, despite its latitudinal proximity to London, suffered harsh winters, the mystery of North America's climate deepened. Even worse, many boosters' sanguine claims about the invigorating benefits of North America's air on human bodies proved illusory as disease, famine, and starvation became fixtures of colonial life. The ongoing, rapid demographic collapse of American Indian communities also had ominous overtones about the continent's long-term habitability.[2]

North America's puzzling climate attracted sensational, hyperbolic commentary. In 1752, for instance, the Rhode Island minister Reverend James MacSparran penned a series of letters to Irish friends warning against emigration. These letters, published the following year as *America Dissected, Being a Full and True Account of all the American Colonies*, vilified the erratic and deadly colonial climates. MacSparran bemoaned the extreme seasonal temperatures and atmospheric fluctuations that unleashed fierce thunderstorms. 'Our Nearness to the Sun occasions more frequent and loud Claps of Thunder, and sharper Lightning, than you have', he cautioned. 'It is no unusual Thing for Houses, and Stacks of Hay, and Grain, to be burnt; and Men and Cattle are often killed by the sharp Lighting.' Dramatic temperature shifts were also killers. 'We are sometimes frying, and at others freezing', he continued. 'Men often die at their Labour in the Field by *Heat*, so some in Winter are froze to Death with the *Cold*.' Finally and most ominously, MacSparran heaped scorn on the 'bad and unwholesome Air' that was 'destructive to human bodies'.[3]

McSparran was hardly alone. Shortly after the publication of his emigrant's guidebook, the well-known trans-Atlantic controversy about North America's degenerative climates erupted. During the third quarter of the eighteenth century, the eminent French naturalist, Georges-Louis Leclerc, Comte de Buffon, popularized the idea that miasmatic marshes blanketed the continent's surface. In his forty-four-volume *Histoire naturelle*, Buffon insisted that North America, just prior to colonization, had experienced a transcontinental deluge. As a result, long stretches of marshes and swamps, towering forests, impenetrable undergrowth, and a frigid and miasmatic atmosphere predominated. Buffon hypothesized that cold temperatures and the abundance of surface water had a degenerative impact on flora and fauna. The continent allegedly had fewer large quadrupeds than the Old World, minimal biodiversity, sluggish and slothful small animals, and swarms of insects and reptiles, which thrived in marshy and chilly environs. Touting large quadrupeds as a yardstick for nature's vitality, Buffon deplored that camels, dromedaries, giraffes, hippopotamuses, and lions did not populate the New World. The chilly and sodden landscape produced an indolent, sickly, and unimpressive natural order.[4]

European intellectuals disdained undeveloped colonial landscapes, which they periodically attributed to indigenous peoples' alleged indolence. In his *Les Recherches philosophiques sur les Américains*, the Dutch-born Cornelius de Pauw ridiculed North America as an expansive 'fetid and boggy terrain'.[5] In 1767, Adam Ferguson explained that 'the climates of America . . . are observed to differ from those of Europe. There, extensive marshes, great lakes, aged, decayed, and crowded forests, with the other circumstances that mark an uncultivated country . . . replenish the air with heavy and noxious vapours [sic.]'.[6] The Scottish historian William Robertson's popular *History of America* echoed these assessments, emphasizing that 'prodigious marshes overspread the [North American] plains'.[7] Descriptions of North America's sogginess did not entirely lack merit. By the time the 1783 Treaty of Paris ended the American Revolution, the territory of the present-day coterminous United States encompassed 221 million acres of wetlands (11 per cent of its geographic surface), much of it concentrated in eastern, midwestern, and southern states.[8]

North America's marshiness constituted a serious social *and* political problem. Colonists and later Americans vilified wetlands and marshy landscapes as 'waste lands' that inhibited progress and prosperity. Marshes impeded travel, depressed property values, harbored ravenous predators and poisonous reptiles, and afforded sanctuary for hostile Indians and runaway slaves. According to Enlightenment medical, meteorological, and pneumatic chemistry knowledge, the stagnant, pestiferous waters in swamps hastened the decomposition of animal and vegetable matter, releasing miasmas and other bad airs into the adjoining atmosphere. The idea that miasmatic zones encompassed landscapes characterized by stagnant surface waters was a hallmark of early modern climatology. As one scholar recently put it, 'the belief that . . . a swamp, watery climate with impure air shortened the lives of people living there . . . was common in the medical community and among the general public in the early modern period'.[9]

In response to Buffon's grave accusations, Americans framed 'climate' as an infinitely malleable phenomenon whose epidemiological qualities diminished as the scale and scope of landscape modifications increased. In 1769, New Jersey's Edward Antill informed the American Philosophical Society that widespread land drainage, deforestation, and cultivation had improved the air quality. 'Whoever compares the present state of the air', Antill explained, 'with what i[t] was formerly, before the country was opened, cleared and drained, will find that, we are every year fast advancing to that pure and perfect temperament of air'. In a 1770 paper presented to the Philosophical Society, physician Hugh Williamson identified deforestation and wetlands drainage as the principle catalysts for purifying the miasmatic atmosphere. According to Williamson, 'tall timber greatly impedes the circulation of air', making it difficult for 'fresh' ocean breezes to disperse miasmatic concentrations. Before the beginning of European colonization, 'the face of this country was clad with woods, and every valley afforded a swamp or stagnant marsh . . . and [because of] a general exhalation from the surface of ponds and marshes, the air was constantly charged with a gross putrescent fluid'.[10]

The sharpest rebuttal to the degeneracy theory emerged from the pen of Thomas Jefferson. In *Notes on the State of Virginia* (1785), the future president ridiculed Buffon's assertion that the continent's frigid temperatures and abundance of aerial and surface moisture reduced the size and diversity of animal species. Jefferson did not dispute Buffon's claim that North America had 'more waters . . . spread over its surface by nature, and fewer of these drained off by the hand of man', but he demonstrated that the continent's animal species, as a whole, were more impressive in size, stature, and diversity than their Old World counterparts. He also rejected Buffon's contention that North America had a cooler climate than European countries of comparable latitudes. Like his fellow countrymen, Jefferson never doubted that intensive land drainage would dry out the continent's sodden soils, delay the decomposition of sodden animal and vegetable matter, and render unhealthy climates an ephemeral moment during the continent's agricultural development.[11]

The resurgence of yellow fever

In 1793, a fresh outbreak of yellow fever in several cities prompted a bitter medical debate about the disease's etiological origins and the extent of climactic improvement. 'Localists' argued that the disease originated from domestic sources where the putrefaction of animal and vegetable matter discharged miasmas into the adjoining atmosphere. 'Contagionists' retorted that infected people aboard vessels arriving from the West Indies, which was notorious for its debilitated and deadly climate for Europeans, were the source of the outbreaks. As Thomas A. Apel has demonstrated, by 1805 the localists had won the debate, disseminating their ideas through faculty positions, prolific publications, an emphasis on 'common sense', and appeals to divine sovereignty. Nevertheless, in seeking to explain the source of yellow fever, both contagionists and localists reconceptualized 'climate' as a microenvironmental phenomenon that could be easily manipulated by simple feats of civil engineering, ingenuity, and proper governance. In reducing the scale of climate, they also identified surface water distribution as a far more accurate barometer of healthy climates than longitudinal positioning, wind patterns, intensive monoculture, deforestation, and other related factors.[12]

As yellow fever raged up and down the Atlantic seaboard, William Currie, a Philadelphia physician and contagionist, became one of the first Americans to consider the role of surface water in local economies of health. In 1795, he read a paper at the American Philosophical Society arguing that the putrefaction of vegetable and animal matter in marshes triggered a chemical reaction, which diminished atmospheric oxygen levels. Curry cited the eudiometrical experiments of Dutch physician Jacob Van Breda to confirm that the 18:48 ratio of oxygen to 'azote' near marshes was far below the normal atmospheric level of 1:3. 'The causes of the unwholesomeness of low and moist situations', Currie asserted, 'is not owing to any invisible miasmata or noxious effluvia, which issue from the soil and lurk in the air, but to a very different cause, viz. to a deficiency of the oxygenous [sic.] portion of the atmosphere in such situations'. Advancements in medicine, pneumatic chemistry, and human anatomy during the previous two centuries demonstrated oxygen's centrality to cardiovascular and pulmonary functions. Starved of oxygen, bodily functions performed 'imperfectly and languidly'. As Currie explained, oxygen deprivation rendered the 'vessels on the surface of the body powerless, and atonic [sic.]'. The cardiovascular system's weakened state made the body vulnerable to miasmas. 'It appears more than probable', Currie concluded, that 'febrile contagion . . . is rendered virulent and powerful in proportion to the absence or defect of the oxygen and the degree of heat to which the living body has been exposed'.[13]

Boosting oxygen levels around marshes could enhance people's resistance to fevers. By building a system of 'drains, deep trenches, and wells' through marshes that conveyed stagnant surface water into flowing watercourses, people increased oxygen levels. According to Currie, physicians should implore farmers to fill low, miry spots with clay, sand, or lime; set fire to dead and decomposing weeds, grass, and trees; and sow grasses, 'plants of vigorous growth', and vegetables for the purpose of 'replenish[ing] the atmosphere with oxygen'.[14]

Some physicians broke with their profession and blamed atmospheric contamination on excessive land drainage. In December of 1798, for instance, physician Adam Seybert appeared before the American Philosophical Society and described a series of eudiometrical experiments he conducted around Philadelphia marshes in 1796 and 1798. Seybert's measurements demonstrated that marshes hastened the putrefaction of vegetable and animal matter, triggering a chemical reaction that absorbed oxygen and 'destroy[ed] the purity of the atmosphere'.[15] Emphasizing that oxygen-deficient landscapes degraded the health of nearby organisms, he attributed the return of yellow fever to extensive drainage, which contributed to a deadly buildup of oxygen levels. As Seybert put it, 'I consider [marshes] as very necessary to keep the atmosphere in a *proper degree* of purity, for it is not only the impure atmosphere which kills animals, but the too pure also'.[16] If marshes did indeed help maintain an acceptable balance of oxygen, azotic gas, and carbonic acid gas, the rapid deforestation and drainage initially applauded for improving the continent's climates were having the reverse effect.

Seybert's ominous warnings about the perils of too much drainage fell on deaf ears. The continent's abundance of surface waters remained a primary explanation for the onset of fevers. In 1802, Charles Caldwell, a student of Dr. Benjamin Rush and later founder of Louisville University's medical school, published one of the era's defining studies on the relationship between undrained landscapes and fever eruptions. Caldwell identified yellow fever as a symptom of a larger problem related to surface water distribution. Entitled *An Oration on the Causes of the Difference . . . Between the Endemic Disease of the United States of America and those of the Countries of Europe*, Caldwell's booklet juxtaposed nature in Europe and North and South America. Taking a hemispheric perspective, Caldwell argued that the New World's 'diversity of latitudes and climates . . . aid a foundation for a state of things unknown in other quarters of the Globe'. Indeed, he extolled the towering size and majesty of New World forests, mammals, lakes, mountains, rivers, and waterfalls, which dwarfed their European counterparts. According to Caldwell, it was natural, if not predictable, that the western hemisphere's impressive topographical and hydrographic phenomena created fertile conditions for miasmatic climates.[17]

According to Caldwell, the continent's flat topography, abundance of precipitation, suffocating heat, and raging rivers, which seasonally inundated riparian lands, created long tracts of stagnant and diffused surface waters. Stretching from the Atlantic seaboard's coastal salt marshes to the Mississippi River's alluvial floodplains, the United States was home to 'vast factories of this febrile poison'. Since the assortment of elements required for generating miasma – heat, decomposing animal and vegetable matter, and motionless water – dominated the surface, the United States had a plethora of pathogenic climates: 'The super-abundance of *marsh miasma* in the United States, compared with most parts of Europe, surpasses the super-abundance of our summer hearts, and would seem to be more immediately and powerfully influential in the production of our diseases'.[18]

The yellow fever crisis stemmed from Americans' leisureliness in draining the continent. The miasmatic saturation of the atmosphere, Caldwell continued, owed itself to the 'present medial or halfway state of our agricultural improvements'. The climactic destiny of the young rested in the hands of farmers. As Caldwell put it, marshes, flood-prone lands, and other landforms dominated by motionless waters were 'physical imperfections' awaiting the shovel of enterprising men. Once farmers drained the land and reintroduced surface waters into rivers and streams, miasmas and other unwholesome airs disappeared:

> The most unlettered husbandman . . . knows, that [marshes'] hurtful properties depend on its super-abundance of stagnating humidity, and that the only method of rendering it innoxious to health, and useful in agriculture, is to circumscribe and intersect it with a number of ditches sufficient to carry

off its redundant waters . . . To render [marshes] healthy, and productive in an agricultural point of view, their deficiency of natural drains and outlets for their moisture must be supplied by human industry and enterprize.[19]

The rapid pace of economic growth also provided an alternative framework for conceptualizing the yellow fever crises. If North America's swampiness contributed to its disease-prone climates, so, too, did capitalist enterprises. Physicians and civic leaders attacked economic ventures that impounded watercourses and accelerated the decomposition of previously unsubmerged vegetable and animal matter. As early as 1776, for instance, Lionel Chalmers, a Scottish physician living in Charlestown, South Carolina, blamed rice cultivation, indigo extraction, and milling for creating unhealthy climates. In a passage worthy of quoting at length, Chalmers argued:

> In almost every settlement, much land is designedly overflowed, by stopping the water courses with strong banks of earth; whereby *reservoirs* of a good depth and extent are formed, in order to be let into the rice fields, when the plant is of a fit growth for receiving the water; for extracting the dye from the plant which yields indico [sic.]; or for mills of various sorts. And, whenever these collections of water are expended in the above purposes, or they are exhaled by the sun or swept away by winds, such multitudes of fish and reptiles of various kinds perish, that, for a long time after the air is tainted, with the putrid *effluvia* that arise as well from the numberless bodies of animals, which are in the highest state of putreaction [sic.].[20]

Millponds also received heightened scrutiny. By removing water from the hydrological cycle, millpond owners created their own factories of febrile poison. Recognizing the 'injurious effects' of millponds, Currie in 1795 implored farmers to plant rows of rapid-growing trees between their homes and millponds to intercept miasma. Caldwell also identified millponds as 'our most certain and prolific sources of autumnal disease'. Inundating trees, shrubbery, and other grasses, millponds hastened the decomposition of organic matter. Indeed, in some American communities, terrified residents dismantled dams. In January 1799, Elijah Boardman and a group of New Milford, Connecticut, residents tore down Joseph Ruggles's dam across the Housatonic River. Boardman and his co-conspirators alleged that since 1796, when Ruggles elevated the dam ten inches, 300 town residents had fallen sick due to a mysterious fever. Blaming impounded stagnant waters for poisoning New Milford's air, angry residents justified the dam's destruction on the grounds it constituted a public health menace.[21]

Submerged rice fields also elicited concern. Organized in 1804, the Georgia Medical Society urged government intervention to curb wet rice cultivation. 'With a semi-tropical climate, such as ours', the Society's charter explained, 'there could be no worse or more malignant incidental cause of disease than the stagnant water, which remains on a rice field exposed to an ardent summer's sun, and the saturated soil which is next exposed, when the water is drained off'. In 1810, the Georgia Medical Society encouraged the Savannah city council to implement a program aimed at eliminating wet rice cultivation on the outskirts of town. In 1817, the city approved the plan, imposing taxes on city residents to pay nearby rice planters to drain their property and convert to dry rice culture. The program enjoyed limited success, since most cultivators, despite the subsidies, found it too costly to drain their land or convert to dry cultivation. Three years after the commencement of the program, a yellow fever epidemic struck Savannah, killing 907 people. The Georgia Medical Society blamed planters' dilatoriness in converting to dry culture, the slow pace of swamp drainage in the hinterland, and the absence of tree barriers outside of Savannah for the epidemic.[22]

Western climates and febrile maladies

After 1805, yellow fever disappeared from the United States for a dozen years. The yellow fever crisis of the 1790s and early 1800s did not uniformly dictate the course of nineteenth-century American environmental thought, but it forged a set of common assumptions about surface waters' inimical impact on human and animal health, informing public discourse at the same moment tens of thousands of Americans migrated west into the wet prairie Midwestern states and the swampy Mississippi River valley. In a recent essay, the historian Gordon S. Wood wrongly asserted that antebellum Americans' growing participation in the market economy somehow made them forget the climactic lessons of the 1790s and the physical environment in which they lived, farmed, and raised livestock:

> In time, of course, Buffon's charges and the problems of America's climate were largely forgotten. America's bumptious political and cultural environment, especially with the growth of a market economy, tended to overwhelm the geographical environment. The busyness of Americans, their search for the almighty dollar, seemed to dominate everything, making the physical environment appear inconsequential by nature.[23]

Yet the historical evidence does not bear out this argument. Westward expansion intensified Americans' climactic anxieties by bringing more and more of them into contact with undeveloped landscapes characterized by stagnant surface water, seasonal flooding, and wet soils. Indeed, as historians Conevery Bolton Valenčius and Linda Nash have demonstrated, western settlers used their bodies as barometers to evaluate the healthfulness of uncultivated landforms. Specifically beginning in the second quarter of the 1800s, concern about these landscapes shifted from the eastern seaboard to the block of territory sandwiched between the Appalachian Mountains and the Mississippi River (and later California's Central Valley). The shift followed the admission of several new states to the Union, their opening for settlement, the securing of Mississippi River navigation rights, the gradual breakdown of Indian resistance after General Anthony Wayne's 1794 victory at Fallen Timbers, Ohio, and the defeat of Tecumseh's

Indian alliance. In a few short decades following independence, ten states west of the Appalachians entered the union: Kentucky, Tennessee, Ohio, Louisiana, Indiana, Mississippi, Illinois, Alabama, Maine, and Missouri.[24]

As settlers fanned out across the vast wet prairie states of Ohio, Indiana, Illinois, and the swampy states bordering the Lower Mississippi River watershed, the region entered the public imagination as a prolific incubator of miasmatic poisons. Westerners employed a host of terms to describe the febrile illnesses they attributed to western marshes, swamps, and seasonal floods: 'Arkansaw chills', 'autumnal fever', 'bilbous fever', 'black swamp fever', 'intermittent fever', 'malaria', 'remittent fever', 'seasoning', and 'swamp fever'. Modern epidemiologists recognize that endemic malaria spread by the bite of the female *anopheles* mosquito – not miasmas – triggered many of the eruptions of 'shakes', 'fevers', and 'chills and ague' that afflicted settlers living near swamps, bogs, and shallow lakes, the *anopheles*' preferred breeding habitats, during humid summer months. Despite this etiological misunderstanding, westerners had good reason to associate miasmas with death: by mid-century, for instance, malaria killed more people in Illinois than anything else. Annual bouts of 'shakes' and 'fevers' were so prevalent that mid-century Midwesterners accepted them as a routine, though dreaded and despised, part of everyday rural life. As one Pike County, Illinois, resident explained in 1821: 'An illness native in the prairie country was fever and ague. There was burning fever following chills which left the patient so weak he could not work. It came with perfect regularity.'[25]

So unnerved were some western land seekers about succumbing to miasmatic diseases that they uprooted their families and retreated to the east coast. In 1818, Thomas Nuttall happened upon several individuals near Georgetown, Pennsylvania, who left the West 'in search of a situation which might afford them health'. The next year William Faux met half a dozen wagons hauling several families and their belongings east from Missouri to Kentucky since they had been 'scared out of Missouri by sickness'. Journeying to Indiana in 1821, William Forster observed as many people returning east due to 'sickness' as he saw heading west. Government explorer Stephen H. Long, a major in the U.S. Army, also disparaged Ohio's wet prairies as a mosquito-ridden wasteland. After traversing the 24 miles between Fort Wayne and the Indiana–Ohio boundary, Long judged that

> the country is so wet that we scarcely saw an acre of land upon which a settlement could be made. We travelled for a couple of miles with our horses wading through water, sometimes to the girth . . . We attempted to stop and pasture our horses, but this was quite impossible on account of the immense swarms of mosquitoes and horse flies.[26]

As settlers emigrated westward, American farm journals, which exploded in readership and circulation beginning in the 1820s, became a popular forum for farmers, physicians, and political leaders to share environmental knowledge, opinions, and wisdom. From 1819 to 1861, 400 different periodicals sprang up across the country, although many attracted a small readership and quickly went out of business. Despite the outbreak of the Civil War, American publishers printed and distributed sixty farm newspapers. In 1880, circulation exceeded the one-million threshold. Contributors to American farms journals accepted without reservation that febrile maladies owed their existence to miasmas, miasmata, or other corrupt airs. By the 1840s, articles and editorials in the rural press increasingly referred to these airs as 'malaria' (literally translated 'bad air'). Language used to describe the etiological origins of fevers increasingly emphasized atmospheric causation rather than immoderate 'climates'. As the *Cincinnati Weekly Herald and Philanthropist* editorialized in 1844, malaria 'consists in certain invisible effluvia or emanations from the surface of the earth, which were formerly called Marsh Miasmata, but to which it has of late years become fashionable to apply the foreign term Malaria'.[27]

Farm journals offered a printed medium for ordinary Americans to share knowledge about the danger of corrupt airs. In 1837, for instance, 'R. B. J★★★★' expounded on malaria's 'modus operandi' for readers of the Petersburg, Virginia-based *Farmer's Register.* According to R. B. J★★★★, 'the effluvium arising from marshes, is a subtle, highly attenuated and undefined substance, the nature of which is unknown, and by no principles of analysis or synthesis have the medical part of the community been able to arrive at any satisfactory conclusion in relation to the thing itself'. The only thing certain about miasma's physical nature was its uncertainty. 'Theory succeeding theory has been exploded without arriving at any thing [sic.] like certainty about the substance of miasm [sic.] itself.'[28]

Although contributors struggled to identify malaria's chemical structure, they agreed about its primary origins: stagnant, noncirculating waters. Writing in 1823, 'Rusticus' explained to *The American Farmer* readers that inland swamps released 'nothing but the most pestilential miasma, thereby contaminating the otherwise wholesome atmosphere, and spreading disease and death through a whole region of a fine fertile country'. This narrative predominated for half a century as letters, speeches, and medical treatises reprinted in farm journals reinforced the wetland–malaria nexus. In 1870, 'Medical Man' elucidated for *New England Farmer* subscribers how miasma formed. 'By [miasma] is meant the effluvia, exhalations, [etc.], which emanate from vegetable and animal matter while undergoing decomposition' in swamps and marshes.[29]

Swamps' lethality mirrored their unsightliness. In a mid-century speech at Ohio's Lucas and Fulton County fairs, which the *Ohio Cultivator* reprinted, Lewis Lambert argued that 'stagnant' surface waters and 'deadly' marshes 'prevent[ed] the growth of useful vegetation, sending forth their poisonous vapors, their death bearing miasma; spreading disease and destroying human life, decimating our population, and frightening the emigrant to other lands, a stench and curse to the neighborhood, and a sickening scene of disgust to the traveller [sic.]'. Editors warned farmers about the proper positioning of homes, stables, and other farm buildings. In 1820, the Baltimore-based *American Farmer* editorialized that healthy farm families did not live in homes or labor in barnyards downwind from 'wet and watery places, particularly stagnant pools, ponds, mill dams, marshes, swamps, meadows, &c. &c.'. The Chicago-based *Prairie Farmer* warned readers to erect their homes, stables, and other farm buildings on dry soils distant from stagnant waters: 'Never allow any considerations to draw you into a swamp or the vicinity of one, where the sun of an American summer is sure to engender in some form the seeds of disease, if not of death.'[30]

Malaria was no less injurious to livestock. Antebellum farm journals resuscitated Buffon's theories about the degenerative tendencies of North America's environment. In doing so, they posited an environmental history of health and disease that embodied and recalibrated eighteenth-century climatic theories of disease etiology. The revival and popularization of eighteenth-century European theories of environmental degeneration constituted an existential crisis for antebellum farmers, their families, and livestock living in uncultivated, marshy landscapes.

Farm journals argued that unwholesome airs sapped the health, vigor, and vitality of livestock in three ways. First, wetlands promoted the growth of aquatic plants and 'watery succulent herbage' that contained fewer fortifying nutrients and vitamins than grasses raised on dry soils. The *Cultivator* editorialized that livestock's 'loss of flesh' occurred when they grazed on hydrophytic plants, proving their 'want of nutrient'. Second, poor soil drainage stunted the growth of most plant, hay, and grass species. In 1856, E. Woolverton, an Ohio farmer, warned *Genesee Farmer* readers that surface water rendered the soil unproductive. 'The growth of the plants [on wet soils] is retarded – the health of that plant which is to be used for man or beast, is materially injured, and the health of the *consumer* is injured accordingly. Thus by [the] neglect [of drainage], the health, strength, vigor, and even life of plants may be extinguished'. Third, malaria increased livestock's vulnerability to a special class of bovine diseases. According to the *Cultivator*, cattle that inhaled malaria and foraged on nutrient-deficient aquatic plants degenerated into weak, sickly, and scrawny creatures. 'Several diseases of domestic animals, such as "liver-complaint" in cattle, and "rot" in sheep, are known to be connected with the same causes which produce the diseases in man', the journal editorialized in 1849. 'The effects of malaria and watery succulent herbage, in producing the rot, have long been known.' A decade later, D.A.A. Nichols wrote in a St. Louis-based journal that it was unwise to forage sheep near Mississippi Valley swamps since they 'are subject to miasma – my own opinion would be, that it would not be very healthy for sheep or the shepherd, and would hardly be profitable to undertake raising wool in such sections.'[31]

Draining the land constituted the panacea for combating unhealthy airs. Conveying stagnant water off the land, wrote Alabama's N.T. Sorsby in 1849, 'destroys the noxious miasmata that wet soils and decomposing vegetable matter so rapidly generates during the summer and fall, to [farmers'] great annoyance and danger'. Yet Sorsby cautioned that proper drainage involved more than digging drainage ditches or burying U-shaped clay tiles underground to channel water into natural outlets. 'By *Draining*', Sorsby expounded, 'is meant not only the construction of artificial channels for water in wet soils, but also includes the operations of plowing, digging, and working soils reputed *dry*, which effects drainage by opening passages for the descent of water from the superficial to the lower strata.'[32] Farm journals pointed to the reduction of human and livestock fatalities in England's formerly fenny sections as evidence that drainage improved human *and* animal health. In 1849, *The Cultivator* reminded subscribers that:

> the rural population of drained districts in England have often remarked [on] the favorable effects of drainage on the health and improvement of animals, by which losses of stock have been prevented to a great extent . . . As might be expected, the health of sheep and cattle has been benefitted by drainage to an equal or greater degree than that of the human race.'[33]

As an unhygienic landscape capable of killing or sickening people and livestock by the simple virtue of their existence, wetlands demanded immediate drainage. Delay only invited death and misery.

The biggest obstacle to drainage for western communities was the glut of federally owned lands. Unlike the original thirteen colonies, most states carved out of the territory west of the Appalachians entered the Union as public-land states. Federal land ownership rendered the local water management regimes that developed during the colonial period ineffective since federal property could not be taxed by states or communities or included in local commissions of sewers or levee boards. Furthermore, vacant public wetlands, westerners charged, poisoned the atmosphere with miasmas. By the 1830s, communities west of the Appalachian Mountains clamored that Congress had a responsibility to promote surface water removal by paying to drain the public domain, quickly disposing of unsold public lands, or ceding them to the counties or states so they could sell the lands and invest the revenue in building drainage projects. In 1845, the *St. Louis New Era* editorialized that Congress, by leaving large blocks of federal wetlands undeveloped or unsold, was culpable for killing 'hundreds' of Missourians:

> The Great Swamps that extend over many counties in south east Missouri . . . are well worthy of the serious consideration of members of Congress. It is very clear that those extensive swamps have an evil influence on the health of several States, and prevent much rich land from being cultivated and improved. The lives of hundreds of excellent citizens are annually lost, by reasons of the exhalations from these morasses.[34]

Public health fears led many western communities and boosters to look to federal authorities. From 1827 to 1849, unaffiliated southern and western communities attacked Congress's dilatoriness in disposing of publicly owned wetlands and accused federal legislators of imposing financial and medical hardships on their communities. Louisiana was the first state to express outrage. In 1828, angry state legislators demanded that Congress explain why half of the public lands in Louisiana remained unsurveyed and unsold. The 'peculiar' situation inhibited Louisiana's settlement, impeded the production of agricultural commodities, and burdened landowners with 'great losses' and 'inconveniences'.

Forced to build levees on adjacent public lands to protect their own property, riparian plantation owners deserved relief:

> In order to protect their own plantations from inundation, [riparian proprietors] have had to raise and keep in repair, embankments in front of the public lands that lie on the margins of water-courses; to procure the necessary intercourse between the different parts of the State, and to communicate with their home markets, they have been obliged to build bridges and open public roads on those lands, and more than one-half of the whole male population of Louisiana, from sixteen to forty-five years, have, for the last ten years, and at this time do work at least five days in the year to the making and repairing of those roads, bridges, and embankments on the public lands alone.[35]

Within a decade, local communities from other public-land states joined the angry chorus. The small trickle of local petitions to Congress quickly turned into a deluge. During the late 1830s and 1840s, nearly four dozen petitions and resolutions from state legislatures, counties, local drainage conventions, and private citizens poured in to Congress from Arkansas, Florida, Illinois, Indiana, Iowa, Louisiana, Mississippi, Missouri, and Wisconsin. The petitions asked Congress to subsidize drainage in public-land states by ceding its 'worthless' and 'refuse' wetlands so the states or counties could sell them and then invest the revenue in building their own drainage projects. Other petitions favored the creation of a national drainage program, sharp reductions in the price of unsold federal swamplands, or congressional authorization to build drainage works on federal lands.[36]

The revival of eighteenth-century degeneracy theories armed proponents of federal wetlands cession with a powerful rhetorical weapon. It was bad enough, westerners and southerners argued, that the glut of uncultivated federal lands slowed population growth in their sections, but the threat posed to public health by vacant and undeveloped wetlands was even more terrifying. Miasmas and malaria discharged by swamps and overflow lands sickened nearby communities, depressed the values of neighboring plantations and farms, and, in the opinion of Mississippi Sen. Jefferson Davis, rendered adjacent lands 'so unhealthy that lands susceptible of cultivation cannot be occupied'.[37]

The groundswell of discontent spurred Congress into action. During a series of congressional debates in the late 1840s, proponents of federal wetlands cession invoked prevailing medical ideas to frame the discussion from the perspective of public health. Champions of cession demonized swamps and riparian lands subject to flooding as 'a fertile cause of disease'; 'generative of noxious influences which were injurious to human health'; 'prolific source[s] of disease'; 'fruitful promoters of disease and death'; and 'pestilential to the people of the country'.[38] Since the drainage of stagnant surface waters was a matter of life and death, Congress had a moral obligation to act.

In 1849, 1850, and 1860, Congress authorized three 'Swamp Land Acts' that ceded federal 'swamp and overflow' lands to fifteen southern, midwestern, and western states on the condition that they sell those lands and invest the revenue in building drainage and local flood-control works. Envisioning a partnership between the federal and state governments, the Swamp Land Acts intended to ensure that communities in public-land states would not bear responsibility for draining wetlands or building levees. The legislation followed the nineteenth-century pattern of the federal government inconspicuously acting as a 'coordinator' and 'facilitator' to solve pressing, complex problems through intermediaries such as local governments. By subsidizing surface water removal with public land subsidies, Congress kept responsibility for drainage local and allowed communities to develop the appropriate administrative machinery necessary to eliminate the continent's vast factories of febrile poison.[39]

The rampant graft, fraud, and abuse that plagued the subsequent cession of federal wetlands is beyond the scope of this essay. From their inception, the Swamp Land Acts proved an embarrassing and abysmal failure. Rather than selling their federal grants and subsidizing drainage projects, states and local communities used their grants to promote railroad development, construct public buildings, build bridges and roads, and pay for education and benevolent associations. Nevertheless, the original passage of the laws owed itself in large measure to the persistence of eighteenth-century climactic theories of disease, which stigmatized wetlands as the only landscape capable of killing or degenerating human and animal organisms by the simple virtue of their existence.[40]

In a recent historiographical essay, the historian Joyce E. Chaplin argues that early American scholars can do more to integrate climate history into their discipline by revisiting the eighteenth-century dispute about the New World's erratic, sickly, and degenerative climates. Yet the debate itself did far less to frame premodern climactic discourse and environmental thought than the medical community's cocksure response to the 1790s yellow fever crises, the rural press's dissemination of those ideas, and the identification of surface water distribution as the primary barometer of healthy climates and environs. Ultimately, in the pages of antebellum farm journals, the environmental history of disease and health revivified fears about the degenerative tendencies of undeveloped, marshy landscapes. As the primary vehicle for eliminating miasmatic landscapes, preserving the health of farm families and domestic animals, and averting the degeneration of native flora and fauna, drainage eclipsed all other forms of land conservation prior to the rise of post–Civil War conservationist thinking.[41]

The growth of irrigation in the post–Civil War western United States has dominated American water historiography. Yet irrigation and drainage were always flip sides of the same coin, and scholars would do well to revisit the rich and fertile discourse surrounding anthropogenic climate change and drainage in early America. As liminal spaces where water and land intermingled, all classes of wetlands deserved to be drained, cleared, and cultivated. The belief that bodies communicated the healthfulness of landforms elevated drainage into a paramount public policy objective and cultural fetish. Ultimately, this assortment of related ideas did just as much to shape American political institutions and environmental beliefs as irrigation. Economic opportunism and public health concerns reinforced one another as motivations for drainage, even after the discovery of the mosquito vectors for malaria and yellow fever at the turn of the twentieth century. As a result, the quantity of American land improved by drainage dwarfed the amount of irrigated land by the conclusion of World War I, and most of the drainage occurred under the administration of autonomous local governments known as 'drainage districts'. Small wonder that public health fears, largely influenced by premodern climactic theories, played no less a role in shaping attitudes about land management than capitalism and economic opportunism.[42]

Notes

1 Jan Golinski, *British Weather and the Climate of Enlightenment* (Chicago: University of Chicago Press, 2007), 198. In a recent article, Golinski also argues that the string of yellow fever outbreaks beginning in 1793 undercut Americans' narrative of positive anthropogenic climactic change. See Golinski, 'Debating the Atmospheric Constitution: Yellow Fever and the American Climate', *Eighteenth-Century Studies* 49 (Winter, 2016): 149–165. For Yellow Fever in the United States, see K. David Patterson, 'Yellow Fever Epidemics and Mortality in the United States, 1693–1905', *Social Science and Medicine* 34 (April, 1992): 855–859; Gerald N. Grob, *The Deadly Truth: A History of Disease in America* (Cambridge, MA: Harvard University Press, 2002), 102–103. On the theological and scientific debates about yellow fever's etiological origins, particularly between 'contagionists' and 'localists', see Thomas A. Apel, *Feverish Bodies, Enlightened Minds: Science and the Yellow Fever Controversy in the Early American Republic* (Stanford: Stanford University Press, 2016).

2 The best work on ancient and Enlightenment theories of climates remains Clarence Glacken, *Traces on the Rhodian Shore: Nature and Culture in Western Thought from Ancient Times to the End of the Eighteenth Century* (Berkeley: University of California Press, 1967). See also Richard H. Grove, *Green Imperialism: Colonial Expansion, Tropical Island Edens and the Origins of Environmentalism, 1600–1860* (Cambridge: Cambridge University Press, 1995); James Rodger Fleming, *Historical Perspectives on Climate Change* (Oxford: Oxford University Press, 1998). On climactic discourse resulting from early American colonization and settlement, see Vladimir Janković, 'Climates as Commodities: Jean Pierre Purry and the Modelling of the Best Climate on Earth', *Studies in the History and Philosophy of Science Part B* 41 (September, 2010): 201–207; Brant Vogel, 'The Letter from Dublin: Climate Change, Colonialism, and the Royal Society in the Seventeenth Century', *Osiris* 26 (2011): 111–128; Sam White, 'Unpuzzling American Climate: New World Experience and the Foundations of a New Science', *Isis* 106 (September, 2015): 544–566; Anya Zilberstein, *A Temperate Empire: Making Climate Change in Early America* (Oxford: Oxford University Press, 2016). On the American climate and atmosphere, see Karen Ordahl Kupperman, 'The Puzzle of the American Climate in the Early Colonial Period', *American Historical Review* 87 (December, 1982): 1262–1289; Kupperman, 'Fear of Hot Climates in the Anglo-American Colonial Experience', *William and Mary Quarterly* 41 (April, 1984): 213–240; William B. Meyer, *Americans and Their Weather* (Oxford: Oxford University Press, 2000), 17–42; Andrew Wear, 'Place, Health, and Disease: The *Airs, Waters, Places* Tradition in Early Modern England and North America', *Journal of Medieval and Early Modern Studies* 38 (Fall, 2008): 443–465. For an insightful discussion of early Americans' anxieties about their environment, see Gordon Wood, *The Rising Glory of America, 1760–1820* (rev. ed., Boston: Northeastern University Press, 1990), 15–20. On the atmospheric wonders of the New World, see James Delbourgo, *A Most Amazing Scene of Wonders: Electricity and Enlightenment in Early America* (Cambridge, MA: Harvard University Press, 2006), 50–51, 237–238. There is a vast literature on the demographic collapse of American Indian communities. Standard works include: Alfred W. Crosby, Jr., *Ecological Imperialism: The*

Biological Expansion of Europe, 900–1900 (2nd ed., Cambridge: Cambridge University Press, 2004); Crosby, 'Virgin Soil Epidemics as a Factor in the Aboriginal Depopulation of America', *William and Mary Quarterly*, 3rd series 33 (April, 1976): 289–299; Jared Diamond, *Guns, Germs, and Steel: The Fate of Human Societies* (New York: W.W. Norton & Company, 2005); and Russell Thornton, *American Indian Holocaust and Survival: A Population History since 1492* (Norman: University of Oklahoma Press, 1987).

3 James M[a]cSparran, *America Dissected, being a Full and True Account of All the American Colonies: Shewing, the Intemperance of the Climates; Excessive Heat and Cold, and Sudden Violent Changes of Weather; Terrible and Mischievous Thunder and Lightning; Bad and Unwholesome Air, Destructive to Human Bodies; Badness of Money; Danger from Enemies; But, above All, the Danger to the Souls of the Poor People That Remove Thither, from the Multifarious Wicked and Pestilent Heresies That Prevail in Those Parts* (Dublin: S. Powell, 1753), 39.

4 Antonello Gerbi, *The Dispute of the New World: The History of a Polemic, 1750–1900*, trans. Jeremy Moyle (Pittsburgh: University of Pittsburgh Press, 1973), 3–14; Glacken, *Traces on the Rhodian Shore*, 680.

5 Quoted in Philippe Roger, *The American Enemy: A Story of French Anti-Americanism*, trans. Sharon Bowman (Chicago: University of Chicago Press, 2005), 7.

6 Quoted in Ralph N. Miller, 'American Nationalism as a Theory of Nature', *William and Mary Quarterly*, 3rd series, 12 (January, 1955): 79.

7 Quoted in Gilbert Chinard, 'Eighteenth Century Theories on America as a Human Habitat', *Proceedings of the American Philosophical Society* 91 (February, 1947): 38.

8 For the quantity of American wetlands during the early republic, see Thomas E. Dahl, *Wetlands Losses in the United States 1780's to 1980's* (Washington: U.S. Department of the Interior, U.S. Fish and Wildlife Service, 1990), 1, 6.

9 Lydia Barnett, 'The Theology of Climate Change: Sin as Agency in the Enlightenment's Anthropocene', *Environmental History* 20 (February, 2015): 226. On American wetlands, see Jeffrey K. Stine, *America's Forested Wetlands: From Wasteland to Valued Resource* (Durham, NC: Forest History Society, 2008); Hugh Prince, *Wetlands and the American Midwest: A Historical Geography of Changing Attitudes* (Chicago: Chicago University Press, 1997); Ann Vileisis, *Discovering the Unknown Landscape: A History of America's Wetlands* (Washington: Island Press, 1997); William B. Meyer, 'From Past to Present: A Historical Perspective on Wetlands,' in *Wetlands*, ed. Sharon L. Spray and Karen L. McGlothlin (Lanham: Rowman & Littlefield Publishers, Inc., 2004), 84–100. On swamps as 'wastelands', see Vittoria Di Palma, *Wasteland: A History* (New Haven: Yale University Press, 2014), 84–127.

10 Edward Antill, 'An Essay on the Cultivation of the Vine, and the Making and Preserving of Wine, Suited to the Different Climates in North-America', *Transactions of the American Philosophical Society* 1 (January, 1769): 120; Hugh Williamson, 'An Attempt to Account for the Change of Climate, Which Has been Observed in the Middle Colonies in North-America', *Transactions of the American Philosophical Society* 1 (January, 1770): 272–280, quotes at 278, 279, and 280.

Emphases in original. Lionel Chalmers, *An Account of the Weather and Disease of South-Carolina* (London: Edward and Charles Dilly, 1776): 1, 6.

11 Thomas Jefferson, *Notes on the State ofVirginia*, ed. William Peden (Chapel Hill: University of North Carolina Press, 1982), quote at 47, 80. On Jefferson's rebuttal of Buffon's theory of degeneration, see Claudine Colden, *The Fate of the Mammoth: Fossils, Myth and History*, trans. William Rodarmor (Chicago: University of Chicago Press, 2002), chapter 5; Gerbi, *The Dispute of the New World*, 252–258; Charles A. Miller, *Jefferson and Nature: An Interpretation* (Baltimore, MD: Johns Hopkins University Press, 1988), 61–63; Anthony Wilson, *Shadow and Shelter: The Swamp in Southern Culture* (Jackson: University Press of Mississippi, 2005), 7–8; Lee Alan Dugatkin, *Mr. Jefferson and the Giant Moose: Natural History in Early America* (Chicago: University of Chicago Press, 2009).

12 In the early 1800s, as Michael Brown has also argued, Great Britain witnessed 'a subtle but highly significant shift in emphasis away from the idea that epidemic diseases were produced by climatic conditions and toward the idea that they derived from specific

local corruption of the atmosphere caused by putrefying matter.' Michael Brown, 'From Foetid Air to Filth: The Cultural Transformation of British Epidemiological Thought, ca. 1780–1848', *Bulletin of the History of Medicine* 82 (Fall, 2008): 518. On the 'localists' versus 'contagionists'' controversy, see Apel, *Feverish Bodies*, 3, 7–8, 16, 145–150.

13 William Currie, 'An Enquiry into the Causes of the Insalubrity of Flat and Marshy Situations: And Directions for Preventing or Correcting the Effects Thereof', *Transactions of the American Philosophical Society* 4 (1799): 128, 135, 138–139. On Van Breda's eudiometrical experimentation, see Huib J. Zuidervaart, 'An Eighteenth-Century Medical-Meteorological Society in the Netherlands, Instrumentation and Quantification. Part 2', *British Journal for the History of Science* 39 (March, 2006): 55–58.

14 Currie, *Transactions of the American*, 141–142.

15 Adam Seybert, 'Experiments and Observations, on the Atmosphere of Marshes', *Transactions of the American Philosophical Society* 4 (1799): 428.

16 Emphasis in original. Ibid., 429.

17 Charles Caldwell, *An Oration on the Causes of the Difference, in Point of Frequency and Force, between the Endemic Diseases of the United States of America, and Those of the Countries of Europe, Delivered, By Appointment, to the "Philadelphia Medical Society," on the Fifth Day of February, 1802* (Philadelphia: T. and William Bradford, 1802), 5–10, quote at 6.

18 Ibid., 13–19, quotes at 14.

19 Ibid., 15, 19, 22.

20 Emphases in original. Chalmers, *An Account of the Weather*, 1, 6.

21 Currie, *Transactions of the American*, 142; Caldwell, *An Oration on the Causes*, 25–26; Oliver Wendell Holmes, *Boylston Prize Dissertations for the Years 1836 and 1837* (Boston: Charles C. Little and James Brown, 1838), 55–56.

22 Quoted in Megan Kate Nelson, 'The Landscape of Disease: Swamps and Medical Discourse in the American Southeast, 1800–1880', *Mississippi Quarterly* 55 (Fall, 2002): 557–558.

23 Gordon S. Wood, 'Environmental Hazards, Eighteenth-Century Style', in *Old World, New World: American and Europe in the Age of Jefferson*, ed. Leonard J. Sadosky, Peter Nicolaisen, Peter S. Onuf, and Andrew O'Shaughnessy (Charlottesville: University of Virginia Press, 2010), 28.

24 Conevery Bolton Valen ius, *The Health of the Country: How American Settlers Understood Themselves and Their Land* (New York: Basic Books, 2002); Linda Nash, *Inescapable Ecologies: A History of Environment, Disease, and Knowledge* (Berkeley: University of California Press, 2006). Two studies that also probe environment-body linkages are Joyce E. Chaplin, *Subject Matter: Technology, the Body, and Science on the Anglo-American Frontier, 1500–1676* (Cambridge, MA: Harvard University Press, 2001); and Kathleen M. Brown, *Foul Bodies: Cleanliness in Early America* (New Haven: Yale University Press, 2009).

25 Quoted in Lucinda McCray Beier, *Health Culture in the Heartland, 1880–1980: An Oral History* (Urbana: University of Illinois Press, 2008), 4. Also see Michael A. Urban, 'An Uninhabited Waste: Transforming the Grand Prairie in Nineteenth Century Illinois, USA', *Journal of Historical Geography* 31 (October, 2005): 647–665; Peter C. Baldwin, 'How Night Air Became Good Air, 1776–1930', *Environmental History* 8 (July, 2003): 412–429; and Lillian Krueger, 'Motherhood on the Wisconsin Frontier (II)', *Wisconsin Magazine of History* 29 (March, 1946): 333–335. On malaria in the United States, see Margaret Humphreys, *Malaria: Poverty, Race, and Public Health in the United States* (Baltimore, MD: Johns Hopkins University Press, 2001).

26 Ronald Reese, 'Under the Weather: Climate and Disease, 1700–1900', *History Today* 46 (January, 1996): 39; Michael Williams, 'Agricultural Impacts in Temperate Wetlands', in *Wetlands: A Threatened Landscape*, ed. Michael Williams (Oxford: Basil Blackwell, 1990), 204.

27 'An Important Article to Sufferers of Fever and Ague', *Cincinnati Weekly Herald and Philanthropist*, November 27, 1844. On the nineteenth-century American rural press, see Lake Douglas, '"To Improve the Soil and the Mind": Content and Context of

Nineteenth-Century Agricultural Literature', *Landscape Journal* 25 (2006): 69–76; Donald B. Marti, 'Agricultural Journalism and the Diffusion of Knowledge: The First Half-Century in America', *Agricultural History* 54 (January, 1980): 28–37.

28 R.B. J****, 'Remarks on Marsh Effluvia', *Farmer's Register* 5 (July 1, 1837): 142.

29 Rusticus, 'Inland Swamps', *American Farmer* 5 (June 20, 1823): 100; A Medical Man, 'Medical Tropics', *New England Farmer* 4 (July, 1870): 330.

30 'The Health of the Planters and Farmers, and of Their Families', *American Farmer* 2 (June 23, 1820): 100; 'Drainage of Wet Lands', *Ohio Cultivator* 7 (April 1, 1851): 103; 'Position of Farm Houses', *Prairie Farmer* 9 (October, 1849): 316.

31 Many antebellum physicians argued that animals were highly vulnerable to malaria. As one practitioner put it, 'it has been observed by most writers upon the subject, that domestic animals are the first to suffer from the presence of Malaria'. See Alfred T. Magill, *Three Lectures on the Origin and Properties of Malaria or Marsh Miasma, with the Best Means of Preventing its Formation and of Obviating Its Effects on the Human Constitution, when this Cannot be Done*, vol. 6 (Charlottesville: Watson & Tompkins, 1834), 6–7; 'Drainage of Land', *Cultivator* 6 (June, 1849): 175; E. Woolverton, 'Draining Prairie Land', *Genesee Farmer* 17 (May, 1856): 141; D.A.A. Nichols, 'Wool Growing in Missouri', *Valley Farmer* 12 (April, 1860): 112.

32 Emphasis in original. N.T. Sorsby, 'Wet Soils and Their Drainage. No. II', *Southern Cultivator* 7 (May, 1849): 65.

33 'Drainage of Land', *Cultivator* (June, 1849): 175. Several authors cited improvements in English public health following the drainage of fenlands. In 1847, the *Ohio Cultivator* editorialized that 'tracts of land in England, which were liable to fevers and agues, and consumptions, by a complete drainage have become salubrious, and are now upon an average standard of longevity with other parts of the country'. See 'On Draining Lands', *Ohio Cultivator* 3 (January 1, 1847): 2.

34 Quoted in *Jeffersonian Republican* (New Orleans, Louisiana), November 25, 1845. Early in the eighteenth century, colonial legislatures formally adopted 'commissions of sewers', English institutions of medieval origin, to initiate, manage, and fund drainage and water projects. Commissions of sewers empowered individuals to regulate the flow of water into meadow fields or remove obstructions from watercourses under the supervision of the colonial governor or legislature. Colonial laws and customs that regulated the control of water were the seeds from which the United States' localized system of land drainage and flood control grew and flourished. By the close of the nineteenth century, drainage districts, which were a more sophisticated version of commissions of sewers, proliferated throughout the Midwest and South, transforming nature as they transformed the relationship between people, local communities, and local and federal governments. See, for instance, John F. Hart, 'Colonial Land Use Law and Its Significance for Modern Takings Doctrine', *Harvard Law Review* 109 (April, 1996): 1252–1300; and Anthony E. Carlson, '"Drain the Swamps for Health and Home": Wetlands Drainage, Land Conservation, and National Water Policy, 1850–1917', (PhD dissertation, University of Oklahoma, Norman, OK, 2010), 83–89. On colonial water management practices and consequences, see Brian Donahue, *The Great Meadow: Farmers and the Land in Colonial Concord* (New Haven: Yale University Press, 2004), 63, 94, 186–189; Christopher L. Pastore, *Between Land and Sea: The Atlantic Coast and the Transformation of New England* (Cambridge, MA: Harvard University Press, 2014), 34–39. For a good overview of drainage districts in the United States, see Mary R. McCorvie and Christopher L. Lant, 'Drainage District Formation and the Loss of Midwestern Wetlands, 1850–1930', *Agricultural History* 67 (Fall, 1993): 13–39.

35 *Senate Document 10*, 20th Congress, 2nd Session, 1828, 1.

36 Carlson, "Drain the Swamps for Health and Home", 112–113.

37 *Congressional Globe* (hereafter *CG*), 31st Congress, 1st Session, June 13, 1850, 1192.

38 The quotes are respectively from Sen. Borland (Arkansas), Sen. Henry S. Foote (Mississippi), Sen. Borland, Rep. James B. Bowlin (Missouri), and Sen. Benton (Missouri). For a sampling of the references in the Senate and House of Representatives to the association between

wetlands and disease, see *CG*, 30th Congress, 1st Session, August 8, 1848, 30th Congress, 2nd Session, February 24, 1849, 592, February 26, 1849, 594, 31st Congress, 1st Session, June 13, 1850, 1191–2, September 16, 1850, 1826, and September 17, 1850, 1848–9.

39 The Swamp Land Acts applied to the following states: Arkansas, Alabama, California, Florida, Illinois, Indiana, Iowa, Louisiana, Michigan, Minnesota, Mississippi, Missouri, Ohio, Oregon, and Wisconsin. On the federal government as a coordinator and facilitator, see Brian Balogh, *The Associational State: American Governance in the Twentieth Century* (Philadelphia: University of Pennsylvania Press, 2015), 26, 28, 30.

40 There is a voluminous literature on the background and administrative failures of the Swamp Land Acts. See, for instance, Paul Wallace Gates, *History of Public Land Law Development* (Washington: Public Land Law Review Commission, 1968), chap. 13; Karen M. O'Neill, *Rivers by Design: State Power and the Origins of U.S. Flood Control* (Durham: Duke University Press, 2006), 48–53; Prince, *Wetlands and the American Midwest*, 140–148; Vileisis, *Discovering the Unknown Landscape*, 72–91; Martin Reuss, *Designing the Bayous: The Control of Water in the Atchafalaya Basin, 1800–1995* (College Station, TX: Texas A&M University Press, 2004), 40–47.

41 Joyce E. Chaplin, 'Ogres and Omnivores: Early American Historians and Climate History', *William and Mary Quarterly* 72 (January, 2015): 25–32.

42 In 1920, the year the federal government published its first drainage census, the 53 million acres of farmland improved by drainage far exceeded the 19 million acres improved by irrigation. Anthony E. Carlson, 'The Other Kind of Reclamation: Wetlands Drainage and National Water Policy, 1902–1912', *Agricultural History* 84 (Fall, 2010): 453. On drainage districts, see McCorvie and Lant, 'Drainage District Formation and the Loss of Midwestern Wetlands'.

Bibliography

Ackroyd, Peter, *Thames: Sacred River* (London: Vintage, 2008)

Albala, Ken, *Eating Right in the Renaissance* (Berkeley: University of California Press, 2002)

Alberro, Solange, *De gachupín al criollo, O cómo los españoles de México dejaron de serlo* (Mexico City: El Colegio de México, 1992)

Alsop, J.D., 'Ethics in the Marketplace: Gerrard Winstanley's London Bankruptcy, 1643', *Journal of British Studies* 28, no. 2 (1989)

Alsop, J.D., and J.C. Davis, 'Winstanley, Gerrard, (bap. 1609, d. 1676)', in *Oxford Dictionary of National Biography*, ed, H.C.G. Matthew and Brian Harrison (60 vols; Oxford: Oxford University Press, 2004), lix

Altmann, Alexander, 'Judah Halevi's Theory of Climates', *Aleph* 5 (2005)

Andrews, Thomas G., 'Work, Nature, and History: A Single Question, That Once Moved Like Light', in *The Oxford Handbook of Environmental History*, ed. Andrew C. Isenberg (Oxford: Oxford University Press, 2014)

Apel, Thomas A., *Feverish Bodies, Enlightened Minds: Science and the Yellow Fever Controversy in the Early American Republic* (Stanford: Stanford University Press, 2016)

Appuhn, Karl, *A Forest on the Sea: Environmental Expertise in Renaissance Venice* (Baltimore: Johns Hopkins University Press, 2009)

Archer, Ian, *The Pursuit of Stability: Social Relations in Elizabethan London* (Cambridge: Cambridge University Press, 1991)

Ariès, Philippe, *L'Homme devant la mort* (Paris: Seuil, 1977)

Arnold, David, *The Problem of Nature: Environment, Culture and European Expansion* (Oxford: Blackwell, 1996)

Arnold, Ellen F., *Negotiating the Landscape: Environment and Monastic Identity in the Medieval Ardennes* (Philadelphia: University of Pennsylvania Press, 2012)

Arnoux, Mathieu, *Le temps des laboureurs. Travail, ordre social et croissance en Europe (XIe–XIVe siècle)* (Paris: Albin Michel, 2012)

Aroua, Nadjet, 'Les marais de la Mitidja, enjeux et moteur de la colonisation en Algérie', in *Les zones humides méditerranéennes hier et aujourd'hui / Le zone umide mediterranee ieri e oggi*, ed. Magalie Franchomme, Christine Labeur, Daria Quatrida, and Remy Simonetti (Padua: Padova University Press, 2014)

Avis, Paul, *Foundations of Modern Historical Thought: Machiavelli to Vico* (Dover, NH: Croom Helm, 1986)

Baldwin, Peter C., 'How Night Air Became Good Air, 1776–1930', *Environmental History* 8 (2003)

Balogh, Brian, *The Associational State: American Governance in the Twentieth Century* (Philadelphia: University of Pennsylvania Press, 2015)

Bankoff, Greg, 'Winds of Colonisation: The Meteorological Contours of Spain's Imperium in the Pacific 1521–1898', *Environment and History* 12, no. 1 (2006)

Barca, Stefania, *Enclosing Water: Nature and Political Economy in a Mediterranean Valley, 1796–1916* (Cambridge: White Horse Press, 2010)

Barnett, Lydia, 'The Theology of Climate Change: Sin as Agency in the Enlightenment's Anthropocene', *Environmental History* 20 (2015)

Bauer, Ralph, and José Antonio Mazzotti, 'Introduction', in *Creole Subjects in the Colonial Americas*, ed. Ralph Bauer and José Antonio Mazzotti (Chapel Hill: University of North Carolina Press, 2009)

Behringer, Wolfgang, 'Weather, Hunger, Fear: Origins of the European Witch Hunts in Climate, Society, and Mentality', *German History* 13 (1995)

Behringer, Wolfgang, *A Cultural History of Climate*, trans. Patrick Camiller (Cambridge: Polity Press, 2010)

Beier, Lucinda McCray, *Health Culture in the Heartland, 1880–1980: An Oral History* (Urbana: University of Illinois Press, 2008)

Benn, T., *Gerrard Winstanley: A Common Treasury* (London: Verso, 2011)

Benton, Lauren, *A Search for Sovereignty: Law and Geography in European Empires, 1450–1900* (Cambridge: Cambridge University Press, 2010)

Benton, T., *Natural Relations: Ecology, Animal Rights and Social Justice* (London: Verso, 1993)

Bergthaller, Hannes, Rob Emmett, Adeline Johns-Putra, Agnes Kneitz, Susanna Lidström, Shane McCorristine, Isabel Pérez Ramos, Dana Phillips, Kate Rigby, and Libby Robin, 'Mapping Common Ground: Ecocriticism, Environmental History, and the Environmental Humanities', *Environmental Humanities* 5 (2014)

Bernáldez, Andrés, *Historia de los reyes católicos D. Fernando y Doña Isabel (c. 1500)* (2 vols; Seville: Imprenta que fué de J.M. Geofrin, 1870)

Binnema, Theodore, *Enlightened Zeal: The Hudson's Bay Company and Scientific Networks, 1670–1870* (Toronto: University of Toronto Press, 2014)

Blair, Ann, *The Theater of Nature: Jean Bodin and Renaissance Science* (Princeton: Princeton University Press, 1997)

Blomley, N., 'Making Private Property: Enclosure, Common Right and the Work of Hedges', *Rural History* 18, no. 1 (2007)

Bodin, Jean, *Les Six livres de la République* (Paris: Jacques Du Puys, 1576)

Bohstedt, John, *The Politics of Provisions: Food Riots, Moral Economy, and Market Transition in England, c. 1500 1850* (Farnham: Ashgate, 2010)

Boia, Lucian, *The Weather in the Imagination*, trans. Roger Leverdier (London: Reaktion Books, 2005)

Boissière, Jean, 'Population et économie du bois dans la France moderne. Contribution à l'étude des milieux forestiers entre Paris et le Morvan au dernier siècle de l'Ancien régime (vers 1685-vers 1790)' (PhD thesis, Université Paris 1, Paris, 1993)

Bolster, W. Jeffrey, 'Opportunities for Marine Environmental History', *Environmental History* 11, no. 3 (2006)

Bonneuil, Christopher, and Jean-Baptiste Fressoz, *The Shock of the Anthropocene: The Earth, History, and Us* (New York: Verso, 2016)

Boucher, Jacqueline, 'L'incarcération de Jean Bodin pendant la troisième guerre de religion', *Nouvelle revue du seizième siècle* 1 (1983)

Bourde, André, *Agronomie et agronomes en France au XVIIIe siècle* (Paris: SEVPEN, 1967)

Brace, L., *The Politics of Property* (Edinburgh: Edinburgh University Press, 2004)

Braddick, Michael, *God's Fury England's Fire: A New History of England's Civil Wars* (London: Penguin, 2008)

Bravo, Michael, 'Geographies of Exploration and Improvement: William Scoresby and Arctic Whaling, 1782–1822', *Journal of Historical Geography* 32 (2006)

The British Navy and the Use of Naval Power in the Eighteenth Century, ed. Jeremy Black and Philip Woodfine (Atlantic Highlands, NJ: Humanities Press International, 1989)

Brogan, Stephen, *The Royal Touch in Early Modern England: Politics, Medicine, and Sin* (Woodbridge: Boydell for the Royal Historical Society, 2015)

Brooke, John L., *Climate Change and the Course of Global History: A Rough Journey* (Cambridge: Cambridge University Press, 2014)

Brown, Kathleen M., *Foul Bodies: Cleanliness in Early America* (New Haven: Yale University Press, 2009)

Brown, Michael, 'From Foetid Air to Filth: The Cultural Transformation of British Epidemiological Thought, ca. 1780–1848', *Bulletin of the History of Medicine* 82 (2008)

Büntgen, Ulf, and Lena Hellmann, 'The Little Ice Age in Scientific Perspective: Cold Spells and Caveats', *Journal of Interdisciplinary History* 44, no. 3 (2014)

Campbell, Bruce M.S., 'Nature as Historical Protagonist: Environment and Society in Pre-Industrial England', *The Economic History Review* 63, no. 2 (2010)

Candiani, Vera S., 'The Desagüe Reconsidered: Environmental Dimensions of Class Conflict in Colonial Mexico', *Hispanic American Historical Review* 92, no. 1 (2008)

Candiani, Vera S., *Dreaming of Dry Land: Environmental Transformation in Colonial Mexico City* (Stanford: Stanford University Press, 2014)

Cañizares Esguerra, Jorge, 'New Worlds, New Stars: Patriotic Astrology and the Invention of Indian and Creole Bodies in Colonial Spanish America, 1600–1650', *American Historical Review* 104, no. 1 (1999)

Cañizares Esguerra, Jorge, *Nature, Empire, and Nation: Explorations of the History of Science in the Iberian World* (Stanford: Stanford University Press, 2006)

Cañizares Esguerra, Jorge, 'Demons, Stars and the Imagination: The Early Modern Body in the Tropics', in *The Origins of Racism in the West*, ed. Miriam Eliav-Feldon, Benjamin Isaac, and Joseph Ziegler (Cambridge: Cambridge University Press, 2009)

Capp, Bernard, *Astrology and the Popular Press: English Almanacs, 1500–1800* (London: Faber and Faber, 1979)

Carey, Mark, 'The History of Ice: How Glaciers Became an Endangered Species', *Environmental History* 12, no. 3 (2007)

Carlson, Anthony E., '"Drain the Swamps for Health and Home": Wetlands Drainage, Land Conservation, and National Water Policy, 1850–1917' (PhD dissertation, University of Oklahoma, Norman, OK, 2010)

Carlson, Anthony E., 'The Other Kind of Reclamation: Wetlands Drainage and National Water Policy, 1902–1912', *Agricultural History* 84 (2010)

Carney, Judith, and Richard Rosomoff, *In the Shadow of Slavery: Africa's Botanical Legacy in the Atlantic World* (Berkeley: University of California Press, 2009)

Carson, Rachel, *Silent Spring* (London: Penguin, 2000 [1962])

Casey, James, *Early Modern Spain: A Social History* (London: Routledge, 1999)

Cavallo, Sandra, 'Health, Air and Material Culture in the Early Modern Italian Domestic Environment', *Social History of Medicine* 29 (2016), 695–716

Cavallo, Sandra, and Tessa Storey, *Healthy Living in Late Renaissance Italy* (Oxford: Oxford University Press, 2013)

Cavert, William M., *The Smoke of London: Energy and Environment in the Early Modern City* (Cambridge: Cambridge University Press, 2016)

Chabrán, Rafael, 'Medieval Spain', in *Regional Cuisines of Medieval Europe: A Book of Essays*, ed. Melitta Weiss Adamson (New York: Routledge, 2002)

Chakrabarty, Dipesh, 'The Climate of History: Four Theses', *Critical Inquiry* 35, no. 2 (2009)
Chakrabarty, Dipesh, 'Climate and Capital: On Conjoined Histories', *Critical Inquiry* 41, no. 1 (2014)
Chaplin, Joyce E., 'Natural Philosophy and an Early Racial Idiom in North America: Comparing English and Indian Bodies', *William and Mary Quarterly* 54, no. 1 (1997)
Chaplin, Joyce E., *Subject Matter: Technology, the Body, and Science on the Anglo-American Frontier, 1500–1676* (Cambridge, MA: Harvard University Press, 2001)
Chaplin, Joyce E., 'Ogres and Omnivores: Early American Historians and Climate History', *William and Mary Quarterly* 72 (2015)
Charron, Pierre, *De la Sagesse* (vol. 3; Paris: Fayard, 1986)
Chinard, Gilbert, 'Eighteenth Century Theories on America as a Human Habitat', *Proceedings of the American Philosophical Society* 91 (1947)
Cioc, Mark, *The Rhine: An Eco-Biography, 1815–2000* (Seattle: University of Washington Press, 2002)
Ciriacono, Salvatore, *Building on Water: Venice, Holland and the Construction of the European Landscape in Early Modern Times* (Oxford and New York: Berghan Books, 2006)
Coates, Peter, *Nature: Western Attitudes since Ancient Times* (Cambridge: Polity Press, 1998)
Cobo, Bernabé, 'Historia del Nuevo Mundo, c. 1653', in *Obras*, ed. Francisco Mateos (2 vols; Madrid: Biblioteca de Autores Españoles, 1956)
Colden, Claudine, *The Fate of the Mammoth: Fossils, Myth and History*, trans. William Rodarmor (Chicago: University of Chicago Press, 2002)
Colonial Records of Spanish Florida, ed. Jeannette Thurber Connor (2 vols; Deland: Florida Historical Society, 1925)
Conrad, Lawrence, Michael Neve, Vivian Nutton, Roy Porter, and Andrew Wear, *The Western Medical Tradition, 800 BC to AD 1800* (Cambridge: Cambridge University Press, 1995)
Cook, Noble David, *Born to Die, Disease and New World Conquest, 1492–1650* (Cambridge: Cambridge University Press, 1998)
Cornette, Joël, *Le roi de Guerre* (Paris: Payot, 2000)
Corns, T., 'Christopher Hill on Milton, Bunyan and Winstanley', *Prose Studies* 36, no. 3 (2014)
Corns, T., '"I Have Writ, I Have Acted, I Have Peace": The Personal and the Political in the Writings of Winstanley and Some of His Contemporaries', *Prose Studies* 36, no. 1 (2014)
Corns, T., A. Hughes, and D. Loewenstein, 'Introduction', in *The Complete Works of Gerrard Winstanley*, ed. T. Corns, A. Hughes, and D. Loewenstein (2 vols; Oxford: Oxford University Press, 2009)
Cosgrove, Denis, 'Platonism and Practicality: Hydrology, Engineering and Landscape in Sixteenth-Century Venice', in *Water Engineering and Landscape: Water Control and Landscape Transformation in the Modern Period*, ed. Denis Cosgrove and Geoff Petts (London and New York: Bellhaven Press, 1990)
Couzinet, Marie-Dominique, *Histoire et méthode à la Renaissance: une lecture de la Methodus de Jean Bodin* (Paris: Vrin, 1996)
Couzinet, Marie-Dominique, and Jean-François Staszak, 'À quoi sert la "théorie des climats"? Éléments d'une histoire du déterminisme environnemental', *Corpus* 34 (1998)
Cressy, David, *Charles I and the People of England* (Oxford: Oxford University Press, 2015)
Crisis in Europe 1560–1660, ed. Trevor Aston (New York: Doubleday, 1967)
Cronon, William, *Changes in the Land: Indians, Colonists, and the Ecology of New England* (New York: Hill & Wang, 1983)
Crosby, Alfred W., 'Virgin Soil Epidemics as a Factor in the Aboriginal Depopulation of America', *William and Mary Quarterly* 3rd series 33 (1976)
Crosby, Alfred W., *Ecological Imperialism: The Biological Expansion of Europe, 900–1900* (2nd edn; Cambridge: Cambridge University Press, 2004 [1986])

The Cultural Politics of Blood, 1500–1900, ed. Kimberly Anne Coles, Ralph Bauer, Zita Nunes, and Carla L. Peterson (London: Palgrave Macmillan, 2015)

Curare la città: sanità e igiene a Firenze, Roma, Parigi, Londra, Barcellona, ed. Renato Sansa, special issue of *Storia Urbana* 112 (2006)

Dahl, Thomas E., *Wetlands Losses in the United States 1780's to 1980's* (Washington: U.S. Department of the Interior, U.S. Fish and Wildlife Service, 1990)

Darby, H.C., *The Changing Fenland* (Cambridge: Cambridge University Press, 1983)

Darley, Gillian, *John Evelyn: Living for Ingenuity* (New Haven: Yale University Press, 2006)

De Grey, Gary S., *Restoration and Revolution in Britain: Political Culture in the Era of Charles II and the Glorious Revolution* (London: Palgrave Macmillan, 2007)

Degroot, Dagomar, '"Never Such Weather Known in These Seas": Climatic Fluctuations and the Anglo-Dutch Wars of the Seventeenth Century, 1652–1674', *Environment and History* 10, no. 2 (2014)

Delbourgo, James, *A Most Amazing Scene of Wonders: Electricity and Enlightenment in Early America* (Cambridge, MA: Harvard University Press, 2006)

Delgado, James, 'The Continuing Quest: The Lure of the Northwest Passage in the Nineteenth and Twentieth Centuries', in *Arctic Ambitions: Captain Cook and the Northwest Passage*, ed. James K. Barnett and David Nicandri (Anchorage: Anchorage Museum, 2015)

Denis, Vincent, *Une histoire de l'identité, France, 1715–1815* (Seyssel: Champ Vallon, 2002)

De Vries, Jan, 'The Economic Crisis of the Seventeenth Century after Fifty Years', *Journal of Interdisciplinary History* 40, no. 2 (2009)

De Vries, Jan, 'The Crisis of the Seventeenth Century: The Little Ice Age and the Mystery of the "Great Divergence"', *Journal of Interdisciplinary History* 44, no. 3 (2014)

Del Río Moreno, Justo L., 'El cerdo. Historia de un elemento esencial de la cultura castellana en la conquista y colonización de América (siglo XVI)', *Anuario de Estudios Americanos* 53 (1996)

Diamond, Jared, *Guns, Germs, and Steel: The Fate of Human Societies* (New York: W. W. Norton & Company, 2005)

Di Palma, Vittoria, *Wasteland: A History* (New Haven: Yale University Press, 2014)

Donahue, Brian, *The Great Meadow: Farmers and the Land in Colonial Concord* (New Haven: Yale University Press, 2004)

Douglas, Lake, '"To Improve the Soil and the Mind": Content and Context of Nineteenth-Century Agricultural Literature', *Landscape Journal* 25 (2006)

Drayton, Richard, *Nature's Government: Science, Imperial Britain, and the "Improvement" of the World* (New Haven: Yale University Press, 2000)

Duden, Barbara, *The Woman Beneath the Skin: A Doctor's Patients in Eighteenth-Century Germany*, trans. Thomas Dunlap (Cambridge: Harvard University Press, 1991)

Dugatkin, Lee Alan, *Mr. Jefferson and the Giant Moose: Natural History in Early America* (Chicago: University of Chicago Press, 2009)

Earle, Rebecca, *The Body of the Conquistador: Food, Race and the Colonial Experience in Spanish America, 1492–1700* (Cambridge: Cambridge University Press, 2012)

Early Modern Europe: From Crisis to Stability, ed. Philip Benedict and Myron P. Gutmann (Dover, DE: University of Delaware Press, 2005)

Edwards, Paul N., *A Vast Machine: Computer Models, Climate Data, and the Politics of Global Warming* (Cambridge: MIT Press, 2010)

Eltis, David, Philip Morgan, and David Richardson, 'Agency and Diaspora in Atlantic History: Reassessing the African Contribution to Rice Cultivation in the Americas', *American Historical Review* 112, no. 5 (2007)

Esser, Raingard, 'Fear of Water and Floods in the Low Countries', in *Fear in Early Modern Society*, ed. William G. Naphy and Penny Roberts (Manchester: Manchester University Press, 1997)

Febvre, Lucien, and Lionel Bataillon, *A Geographical Introduction to History* (London: Kegan Paul, Trench Trubner & Co., 1925)
Fields, Sherry, *Pestilence and Head Colds: Encountering Illness in Colonial Mexico* (New York: Columbia University Press, 2008)
Fissel, Mary, *Vernacular Bodies: The Politics of Reproduction in Early Modern England* (Oxford: Oxford University Press, 2004)
Fleming, Fergus, *Barrow's Boys* (London: Granta, 1999)
Fleming, James R., *Historical Perspectives on Climate Change* (Oxford: Oxford University Press, 1998)
Floyd-Wilson, Mary, *English Ethnicity and Race in Early Modern Drama* (Cambridge: Cambridge University Press, 2003)
Foucault, Michel, *Naissance de la biopolitique: Cours au Collège de France (1978–1979)* (Paris: Gallimard, 2004)
Foucault, Michel, *Security, Territory, Population: Lectures at the Collège de France 1977–1978*, trans. Graham Burchell (Basingstoke: Palgrave Macmillan, 2007)
Franklin, Julian H., *Jean Bodin and the Sixteenth-Century Revolution in the Methodology of Law and History* (New York: Columbia University Press, 1963)
Franklin, Julian H., *Jean Bodin and the Rise of Absolutist Theory* (Cambridge: Cambridge University Press, 1973)
Fressoz, Jean-Baptiste, *L'apocalypse joyeuse. Une histoire du risque technologique* (Paris: Seuil, 2012)
Fressoz, Jean-Baptiste, and Fabien Locher, 'L'agir humain sur le climat et la naissance de la climatologie historique, XVIIe–XVIIIe siècles', *Revue d'Histoire Moderne et Contemporaine* 1 (2015)
Fulton, Elaine, 'Acts of God: The Confessionalization of Disaster Reformation Europe', in *Historical Disasters in Context: Science, Religion, and Politics*, ed. Andrea Janku, Gerrit J. Schenk, and Franz Mauelshagen (Abingdon: Routledge, 2012)
Fulton, Elaine, and Penny Roberts, 'The Wrath of God: Explanations of Crisis and Natural Disaster in Pre-Modern Europe', in *History at the End of the World: History, Climate Change, and the Possibility of Closure*, ed. Mark Levene, Rob Johnson, and Penny Roberts (Penrith: Humanities-EBooks, 2010)
Galloway, James, and Jonathan Potts, 'Marine Flooding in the Thames Estuary and Tidal River c.1250–1450: Impact and Response', *Area* 40 (2007)
García Ballester, Luis, *La búsqueda de la salud: sanadores y enfermos en la España medieval* (Barcelona: Península, 2001)
Garrioch, David, '1666 and London's Fire History: A Re-Evaluation', *The Historical Journal* 59, no. 2 (2016)
Gascoigne, John, *Science in the Service of Empire: Joseph Banks, the British State, and the Uses of Science in the Age of Revolution* (Cambridge: Cambridge University Press, 1998)
Gascoigne, John, *Joseph Banks and the English Enlightenment: Useful Knowledge and Public Culture* (Cambridge: Cambridge University Press, 2003)
Gates, Paul Wallace, *History of Public Land Law Development* (Washington: Public Land Law Review Commission, 1968)
The General Crisis of the Seventeenth Century, ed. Geoffrey Parker and Lesley M. Smith (2nd edn; London: Routledge, 1997)
Gerbi, Antonello, *The Dispute of the New World: The History of a Polemic, 1750–1900*, trans. Jeremy Moyle (Pittsburgh: University of Pittsburgh Press, 1973)
Glacken, Clarence, *Traces on the Rhodian Shore: Nature and Culture in Western Thought from Ancient Times to the End of the Eighteenth Century* (Berkeley, CA: University of California Press, 1967)

Glimp, David, *Increase and Multiply: Governing Cultural Reproduction in Early Modern England* (Minneapolis: University of Minnesota Press, 2003)

Goldstone, Jack, *Revolution and Rebellion in the Early Modern World* (Berkeley: University of California Press, 1981)

Golinski, Jan, *British Weather and the Climate of Enlightenment* (Chicago: University of Chicago Press, 2007)

Golinski, Jan, 'Debating the Atmospheric Constitution: Yellow Fever and the American Climate', *Eighteenth-Century Studies* 49 (2016)

Gómez, N. Wey, *The Tropics of Empire: Why Columbus Sailed South to the Indies* (Cambridge, MA: The MIT Press, 2008)

Gowland, Angus, 'The Problem of Early Modern Melancholy', *Past & Present* 191 (2006)

Grafe, Regina, 'Economic and Social Trends', in *The Oxford Handbook of Early Modern European History, 1350–1750, vol. 1: Peoples and Place*, ed. Hamish Scott (Oxford: Oxford University Press, 2015)

Grafton, Anthony, *What Was History? The Art of History in Early Modern Europe* (Cambridge: Cambridge University Press, 2007)

Grangel, Luis, *La medicina española renacentista* (Salamanca: Universidad de Salamanca, 1980)

Grob, Gerald N., *The Deadly Truth: A History of Disease in America* (Cambridge, MA: Harvard University Press, 2002)

Grove, Richard, 'Cressey Dymock and the Draining of the Fens: An Early Agricultural Model', *The Geographical Journal* 147, no. 1 (1981)

Grove, Richard, *Green Imperialism: Colonial Expansion, Tropical Island Edens and the Origins of Environmentalism, 1600–1860* (Cambridge: Cambridge University Press, 2003)

Gurney, J., *Brave Community: The Digger Movement in the English Revolution* (Manchester: Manchester University Press, 2007)

Gurney, J., *Gerrard Winstanley: The Digger's Life and Legacy* (London: Pluto Press, 2013)

Gurney, J., 'Gerrard Winstanley and the Concept of Place', *Prose Studies* 36, no. 1 (2014)

Hall, Catherine, 'Introduction: Thinking the Postcolonial, Thinking the Empire', in *Cultures of Empire: A Reader: Colonizers in Britain and the Empire in the Nineteenth and Twentieth Centuries*, ed. Catherine Hall (New York: Taylor & Francis, 2000)

Hallyn, Fernand, *The Poetic Structure of the World: Copernicus and Kepler* (New York: Zone Books, 1990)

Hallyn, Fernand, 'Dialectique et rhétorique devant la "nouvelle science" du XVIe siècle', in *Histoire de la rhétorique dans l'Europe moderne*, ed. Marc Fumaroli (Paris: Presses Universitaires de France, 1999)

Hanawalt, Barbara, and Lisa Kiser, 'Introduction', in *Engaging with Nature: Essays on the Natural World in Medieval and Early Modern Europe*, ed. Barbara Hanawalt and Lisa Kiser (Notre Dame, IN: University of Notre Dame Press, 2008)

Harris, Tim, *Restoration: Charles II and His Kingdoms, 1660–1685* (London: Penguin, 2005)

Harrison, Mark, *Climates and Constitutions: Health, Race, Environment and British Imperialism in India, 1600–1850* (Oxford: Oxford University Press, 1999)

Harrison, P., '"Fill the Earth and Subdue It": Biblical Warrants for Colonization in Seventeenth Century England', *Journal of Religious History* 29, no. 1 (2005)

Hart, John F., 'Colonial Land Use Law and Its Significance for Modern Takings Doctrine', *Harvard Law Review* 109 (1996)

He, Wenkai, 'Public Interest and the Financing of Local Water Control in Qing China, 1750–1850', *Social Science History* 39, no. 3 (2015)

Heilbron, J.L., 'Weighing Imponderables and Other Quantitative Science around 1800', *Historical Studies in the Physical and Biological Sciences* 24, no. 1 (1993)

Heringman, Noah, '"Very Vain Is Science' Proudest Boast": The Resistance to Geological Theory in Early Nineteenth-Century England', *Geological Society of America Memoir* 203 (2009)

Herzog, Tamar, *Defining Nations: Immigrants and Citizens in Early Modern Spain and Spanish America* (New Haven: Yale University Press, 2003)

Hessayon, A., 'Restoring the Garden of Eden in England's Green and Pleasant Land: The Diggers and the Fruits of the Earth', *Journal for the Study of Radicalism* 2, no. 2 (2008)

Hill, C., *The World Turned Upside Down: Radical Ideas during the English Revolution* (London: Penguin, 1972)

Hill, C., 'Introduction', in *Winstanley: The Law of Freedom and Other Writings*, ed. C. Hill (Cambridge: Cambridge University Press, 1983)

Hill, Ruth, 'Entering and Exiting Blackness: A Color Controversy in Eighteenth-Century Spain', *Journal of Spanish Cultural Studies* 10, no. 1 (2009)

Hillgarth, J.N., *The Mirror of Spain, 1500–1700: The Formation of a Myth* (Ann Arbor: University of Michigan Press, 2002)

Hiltner, Ken, *What Else Is Pastoral? Renaissance Literature and the Environment* (Ithaca and London: Cornell University Press, 2011)

Hindle, Steve, 'Dearth and the English Social Revolution: The Harvest Crisis of 1647–50', *Economic History Review* 61 (2008)

Histoire de l'Europe urbaine. I. De l'Antiquité au XVIIIe siècle, ed. Jean-Luc Pinol (Paris: Le Seuil, 2003)

Hobby, E., 'Winstanley, Women and the Family', *Prose Studies* 22, no. 2 (1999)

Hoffmann, Richard, 'Homo et Natura, Home in Natura: Ecological Perspectives on the European Middle Ages', in *Engaging with Nature: Essays on the Natural World in Medieval and Early Modern Europe*, ed. Barbara Hanawalt and Lisa Kiser (Notre Dame, IN: University of Notre Dame Press, 2008)

Hoffmann, Richard, 'Elemental Resources and Aquatic Ecosystems: Medieval Europeans and Their Rivers', in *A History of Water: Series II, vol. 2: Rivers and Society: From Early Civilizations to Modern Times*, ed. Terje Tvedt and Richard Coopey (London: I. B. Tauris, 2010)

Hoffmann, Richard, *An Environmental History of Medieval Europe* (Cambridge: Cambridge University Press, 2013)

Hoffmann, Richard, and Verena Winiwarter, 'Making Land and Water Meet: Cycling of Nutrients between Fields and Ponds in Pre-Modern Europe', *Agricultural History* 84, no. 3 (2010)

Holmes, Clive, 'Drainers and Fenmen: The Problem of Popular Political Consciousness in the Seventeenth Century', in *Order and Disorder in Early Modern England*, ed. Anthony Fletcher and John Stevenson (Cambridge: Cambridge University Press, 1985)

Holmes, Clive, 'Drainage Projects in Elizabethan England: The European Dimension', in *Eau et développement dans l'Europe modern*, ed. Salvatore Ciriacono (Paris: MSH, 2004)

Howkins, A., 'From Diggers to Dongas: The Land in English Radicalism, 1649–2000', *History Workshop Journal* 54 (2002)

Humphreys, Margaret, *Malaria: Poverty, Race, and Public Health in the United States* (Baltimore: Johns Hopkins University Press, 2001)

Hunt, John Dixon, *Greater Perfections: The Practice of Garden Theory* (London: Thames and Hudson, 2000)

Huxley, Margo, 'Geographies of Governmentality', in *Space, Knowledge, Power: Foucault and Geography*, ed. J. Crampton and S. Elden (London: Ashgate, 2007)

Irrigation et drainage dans l'Antiquité, qanâts et canalisation souterraines en Iran, en Égypte et en Grèce, ed. Pierre Briant (Paris: Collège de France/Thotm Edition, 2001)

Janković, Vladimir, 'Climates as Commodities: Jean Pierre Purry and the Modelling of the Best Climate on Earth', *Studies in the History and Philosophy of Science Part B* 41 (2010)

Janković, Vladimir, *Confronting the Climate: British Airs and the Making of Environmental Medicine* (New York: Palgrave Macmillan, 2010)

Jaritz, Gerhard, and Verena Winiwarter, 'On the Perception of Nature in Renaissance Society', in *Nature and Society in Historical Context*, ed. Mikuláš Teich, Roy Porter, and Bo Gustafsson (Cambridge: Cambridge University Press, 1997)

Jenner, Mark, 'The Politics of London Air: John Evelyn's Fumifugium and the Restoration', *Historical Journal* 38, no. 3 (1995)

Jennings, Eric T., *Curing the Colonizers: Hydrotherapy, Climatology, and French Colonial Spas* (Purdue: Duke University Press, 2006)

Joly, Philippe-Louis, 'Jean Bodin', in *Remarques critiques sur le dictionnaire de Bayle* (2 vols; Paris: E. Ganeau, 1752)

Jones, Ryan T., *Empire of Extinction: Russians and the North Pacific's Strange Beasts of the Sea, 1741–1867* (Oxford: Oxford University Press, 2014)

Jonsson, Fredrik A., *Enlightenment's Frontier: The Scottish Highlands and the Origins of Environmentalism* (New Haven: Yale University Press, 2013)

Jonsson, Fredrik A., 'Climate Change and the Retreat of the Atlantic: The Cameralist Context of Pehr Kalm's Voyage to North America, 1748–51', *William and Mary Quarterly* 72, no. 2 (2015)

Jørgensen, Dolly, 'Running Amuck? Urban Swine Management in Late Medieval England', *Agricultural History* 87, no. 4 (2013)

Jørgensen, Dolly, and Sverker Sörlin, 'Making the Action Visible: Making Environments in Northern Landscapes', in *Northscapes: History, Technology, and the Making of Northern Environments*, ed. Dolly Jørgensen and Sverker Sörlin (Vancouver: UBC Press, 2013)

Jouanna, Arlette, *Le Prince absolu. Apogée et déclin de l'imaginaire monarchique* (Paris: Gallimard, 2014)

Jouanna, Jacques, *Hippocrate* (Paris: Fayard, 1992)

Judd, Richard W., *The Untilled Garden: Natural History and the Spirit of Conservation in America, 1740–1840* (Cambridge: Cambridge University Press, 2009)

Kaplan, Steven L., *The Famine Plot Persuasion in Eighteenth-Century France* (Philadelphia: Transactions of the American Philosophical Society, 1982)

Kaplan, Steven L., *Le Pain, le Peuple et le Roi. La bataille du libéralisme sous Louis XV* (Paris: Perrin, 1986)

Kaye, Joel, 'The (Re)Balance of Nature, c.1250–1350', in *Engaging with Nature: Essays on the Natural World in Medieval and Early Modern Europe*, ed. Barbara Hanawalt and Lisa Kiser (Notre Dame, IN: University of Notre Dame Press, 2008)

Kelly, Morgan, and Cormac Ó Gráda. 'The Waning of the Little Ice Age: Climate Change in Early Modern Europe', *Journal of Interdisciplinary History* 44, no. 3 (2014)

Kennedy, G., *Diggers, Levellers and Agrarian Capitalism* (Plymouth: Lexington Books, 2008)

Keohane, Nannerl, *Philosophy and the State in France: The Renaissance to the Enlightenment* (Princeton: Princeton University Press, 1980)

Kesselring, K. J., *The Northern Rebellion of 1569: Faith, Politics, and Protest in Elizabethan England* (London: Palgrave Macmillan, 2007)

Kirshner, Julius, 'Between Nature and Culture: An Opinion of Baldus of Perugia on Venetian Citizenship as Second Nature', *Journal of Medieval and Renaissance Studies* 9 (1979)

Klein, Naomi, *The Shock Doctrine: The Rise of Disaster Capitalism* (New York: Picador, 2007)

Klibansky, Raymond, Erwin Panofsky, and Fritz Saxl, *Saturn and Melancholy: Studies in the History of Natural Philosophy, Religion and Art* (London: Nelson, 1964)

Klima, ed. James R. Fleming and Vladimir Janković, special issue of *Osiris* 26, no. 1 (2011)

Knell, Simon J., 'The Road to Smith: How the Geological Society Came to Possess English Geology', *The Geological Society* (Special Issue: The Making of the Geological Society of London) 317 (2009)

Krueger, Lillian, 'Motherhood on the Wisconsin Frontier (II)', *Wisconsin Magazine of History* 29 (1946)

Kupperman, Karen Ordahl, 'The Puzzle of the American Climate in the Early Colonial Period', *American Historical Review* 87 (1982)

Kupperman, Karen Ordahl, 'Climate and Mastery of the Wilderness in Seventeenth-Century New England', in *Seventeenth-Century New England*, ed. David D. Hall and David Grayson Allen (Boston: Colonial Society of Massachusetts, 1984)

Kupperman, Karen Ordahl, 'Fear of Hot Climates in the Anglo-American Colonial Experience', *William and Mary Quarterly* 41, no. 2 (1984)

Lachiver, Marcel, *Les Années de misère: la famine au temps du Grand Roi, 1680–1720* (Paris: Fayard, 1991)

Lake, Peter, 'From Revisionist to Royalist History: Or, Was Charles I the First Whig Historian?', *Huntington Library Quarterly* 78, no. 4 (2015)

Lake, Peter, 'Post-Reformation Politics, or on Not Looking for the Long-Term Causes of the English Civil War', in *Oxford Handbook to the English Civil War*, ed. Michael Braddick (Oxford: Oxford University Press, 2015)

Lamb, Hubert H., *The Changing Climate: Selected Papers* (London: Methuen, 1966)

Lavallé, Bernard, *Las promesas ambiguas. Criollismo colonial en los Andes* (Lima: Pontífica Universidad Católica del Perú, 1993)

Lestringant, Frank, 'Europe et théorie des climats dans la seconde moitié du XVIe siècle', in *La conscience européenne au XVe et au XVIe siècle. Actes du colloque international organisé à l'Ecole Normale Supérieure des Jeunes Filles (30 sept.–3 oct. 1980)* (Paris: Ecole Normale Supérieure des Jeunes Filles, 1986)

Lestringant, Frank, *Écrire le monde à la Renaissance: quinze études sur Rabelais, Postel, Bodin et la littérature géographique* (Caen: Paradigme, 1993)

Leveau, Philippe, 'Revisiter l'espace et le temps dans le delta du Rhône: archéologie et histoire des zones humides et des milieux deltaïques', *Bulletin Archéologique de Provence* (Supplément) 2 (2003)

Levere, Trevor H., *Science and the Canadian Arctic: A Century of Exploration, 1818–1918* (Cambridge: Cambridge University Press, 1993)

Levine, David, and Keith Wrightson, *The Making of an Industrial Society: Whickham, 1560–1765* (Oxford: Oxford University Press, 1991)

Lindley, Keith, *Fenland Riots and the English Revolution* (London: Heinmann, 1982)

Livi-Bacci, Massimo, *Conquest: The Destruction of the American Indios*, trans. Carl Ipsen (Cambridge: Polity Press, 2008)

Livingstone, David N., 'The Moral Discourse of Climate: Historical Considerations on Race, Place and Virtue', *Journal of Historical Geography* 17, no. 4 (1991)

López Beltrán, Carlos, 'Hippocratic Bodies: Temperament and Castas in Spanish America (1570–1820)', *Journal of Spanish Cultural Studies* 8, no. 2 (2007)

Los cuatro viajes del almirante, Los cuatro viajes del almirante y su testamento, ed. Ignacio Anzoátegui (Madrid: Espasa, 1971)

Lusignan, Serge, *La construction d'une identité universitaire en France* (Paris: Publications de la Sorbonne, 1999)

Lynn, John A., *The Wars of Louis XIV 1667–1714* (New York: Longmann, 1999)

Lyon, Eugene, 'Spain's Sixteenth-Century North American Settlement Attempts: A Neglected Aspect', *Florida Historical Quarterly* 59, no. 1 (1981)

MacCulloch, Diarmaid, and Anthony Fletcher, *Tudor Rebellions* (6th edn; London: Routledge, 2015)

Macdonald, Alan R., and John McCallum, 'The Evidence for Early Seventeenth-Century Climate from Scottish Ecclesiastical Records', *Environment and History* 19, no. 4 (2013)

Maravall, José Antonio, *Estado moderno y mentalidad social, siglos XV a XVII* (2 vols; Madrid: Ediciones de la Revista de Occidente, 1972)

Marcott, Shaun A., Jeremy D. Shakun, Peter U. Clark, and Alan C. Mix, 'A Reconstruction of Regional and Global Temperature for the Past 11,300 Years', *Science* 339, no. 1198 (2013)

Mariscal, George, 'The Figure of the Indiano in Early Modern Spanish Culture', *Journal of Spanish Cultural Studies* 2, no. 1 (2001)

Marti, Donald B., 'Agricultural Journalism and the Diffusion of Knowledge: The First Half-Century in America', *Agricultural History* 54 (1980)

Martin, Craig, *Renaissance Meteorology from Pomponazzi to Descartes* (Baltimore, MD: Johns Hopkins University Press, 2011)

Martínez, María Elena, *Genealogical Fictions: Limpieza de Sangre, Religion, and Gender in Colonial Mexico* (Stanford: Stanford University Press, 2008)

Mattern, Susan, *Galen and the Rhetoric of Healing* (Baltimore: Johns Hopkins University Press, 2008)

Mauelshagen, Franz, 'Disaster and Political Culture in Germany since 1500', in *Natural Disasters, Cultural Responses: Case Studies toward a Global Environmental History*, ed. Christof Mauch and Christian Pfister (Plymouth: Lexington, 2009)

Mauelshagen, Franz, 'Ein neues Klima im 18. Jahrhundert', *Zeitschrift für Kulturwissenschaften* 1 (2016)

Maxwell, Kenneth, *Pombal: Paradox of the Enlightenment* (Cambridge: Cambridge University Press, 1995)

McCombie, Grace, 'The Development of Trinity House and the Guildhall before 1700', in *Newcastle and Gateshead before 1700*, ed. Diana Newton and A.J. Pollard (Chichester: Phillimore, 2009)

McCorvie, Mary R., and Christopher L. Lant, 'Drainage District Formation and the Loss of Midwestern Wetlands, 1850–1930', *Agricultural History* 67 (1993)

McDonagh, B., and C. Griffin, 'Occupy! Historical Geographies of Property, Protest and the Commons, 1500–1850', *Journal of Historical Geography* 53 (2016)

McRae, Andrew, 'Husbandry Manuals and the Language of Agrarian Improvement', in *Culture and Cultivation in Early Modern England: Writing and the Land*, ed. Michael Leslie and Timothy Raylor (Leicester: Leicester University Press, 1992)

McRae, Andrew, *God Speed the Plough: The Representation of Agrarian England, 1500–1660* (Cambridge: Cambridge University Press, 1996)

McRae, Andrew, 'Fluvial Nation: Rivers, Mobility and Poetry in Early Modern England', *English Literary Renaissance* 38, no. 3 (2008)

Medical Cultures of the Early Modern Spanish Empire, ed. John Slater, Maríaluz López-Terrada, and José Pardo-Tomás (Farnham: Ashgate, 2014)

Merchant, Carolyn, *Reinventing Eden: The Fate of Nature in Western Culture* (New York: Routledge, 2003)

Meyer, Torsten, 'Cultivating the Landscape: The Perception and Description of Work in Sixteenth- to Eighteenth-Century German "Household Literature" (*Hausväterliteratur*)', in *The Idea of Work in Europe from Antiquity to Modern Times*, ed. Josef Ehmer and Catharina Lis (Farnham: Ashgate, 2009)

Meyer, William B., *Americans and Their Weather* (Oxford: Oxford University Press, 2000)

Meyer, William B., 'From Past to Present: A Historical Perspective on Wetlands', in *Wetlands*, ed. Sharon L. Spray and Karen L. McGlothlin (Lanham: Rowman & Littlefield Publishers, Inc., 2004)

Miglietti, Sara, 'Amitié, harmonie et paix politique chez Aristote et Jean Bodin', *Astérion* 7 (2010)

Miglietti, Sara, 'Meaning in a Changing Context: Towards an Interdisciplinary Approach to Authorial Revision', *History of European Ideas* 40, no. 4 (2014)

Miglietti, Sara, 'Mastering the Climate: Theories of Environmental Influence in the Long Seventeenth Century' (PhD dissertation, University of Warwick, Coventry, 2016)

Mikhail, Alan, *Nature and Empire in Ottoman Egypt: An Environmental History* (Cambridge: Cambridge University Press, 2013)

Miller, Charles A., *Jefferson and Nature: An Interpretation* (Baltimore: Johns Hopkins University Press, 1988)

Miller, Peter, and Nikolas Rose, *Governing the Present: Administering Economic, Social and Personal Life* (London: Polity Press, 2008)

Miller, Ralph N., 'American Nationalism as a Theory of Nature', *William and Mary Quarterly* 3rd series 12 (1955)

Millstone, Noah, 'Seeing Like a Statesman in Early Stuart England', *Past & Present* 223 (2014)

Miquel, André, *La géographie humaine du monde musulman jusqu'au milieu du XIe siècle* (Paris and La Haye: Mouton, 1975)

Mira Caballos, Esteban, *Indios y mestizos americanos en la España del siglo XVI* (Madrid: Iberoamericana, 2000)

Monahan, W. Gregory, *Year of Sorrows: The Great Famine of 1709 in Lyon* (Columbus: Ohio State University Press, 1993)

Morera, Raphaël, 'Dutch Capital and the Draining of French Wetlands in Arles and Petit Poitou: Environmental Change within an International Network in the Seventeenth Century', *International Review of Social History* 55, Special issue supplement 18 (2010)

Morera, Raphaël, *L'assèchement des marais en France au XVIIe siècle* (Rennes: Presses universitaires de Rennes, 2011)

Morera, Raphaël, 'Mise en valeur des zones humides et associations de gestion. Naissance et affirmation de nouveaux pouvoirs territoriaux (France, XVIe-XVIIIe siècles)', *Siècles* 42 (2015)

Morera, Raphaël, 'La gestion de l'eau dans le Calaisis, les pouvoirs civils et militaires face aux populations locales, fin du XVIIe–début du XVIIIe siècle', in *Aménagement et environnement*, ed. Patrick Fournier and Geneviève Massard-Guilbaud (Rennes: Presses universitaires de Rennes, forthcoming)

Morgan, John, 'Understanding Flooding in Early Modern England', *Journal of Historical Geography* 50 (2015)

Müller, Reimar, 'Montesquieu über Umwelt und Gesellschaft – die Klimatheorie und ihre Folgen', *Sitzungsberichte der Leibniz-Sozietät* 80 (2005)

Narrating the Arctic: A Cultural History of Nordic Scientific Practices, ed. Michael Bravo and Sverker Sörlin (Canton: Watson Publishing International, 2002)

Nash, Linda, *Inescapable Ecologies: A History of Environment, Disease, and Knowledge* (Berkeley: University of California Press, 2006)

Naylor, Simon, 'Log Books and the Law of Storms: Maritime Meteorology and the British Admiralty in the Nineteenth Century', *Isis* 106, no. 4 (2015)

Nef, John, *The Rise of the British Coal Industry* (vol. 1; London: Routledge, 1932)

Nelson, Megan Kate, 'The Landscape of Disease: Swamps and Medical Discourse in the American Southeast, 1800–1880', *Mississippi Quarterly* 55 (2002)

Newson, Linda, and Susie Minchin, 'Diets, Food Supplies and the African Slave Trade in Early Seventeenth-Century Spanish America', *The Americas* 63, no. 4 (2007)

O'Brian, Patrick, *Joseph Banks: A Life* (Chicago: University of Chicago Press, 1987)

Oldroyd, David, 'Maps as Pictures or Diagrams: The Early Development of Geological Maps', *Geological Society of America* Special Paper 502 (2013)

Olson, Richard G., 'Count Rumford, Sir John Leslie, and the Study of the Nature and Propagation of Heat at the Beginning of the Nineteenth Century', *Annals of Science* 26, no. 4 (1970)

O'Neill, Karen M., *Rivers by Design: State Power and the Origins of U.S. Flood Control* (Durham: Duke University Press, 2006)

The Oxford Handbook to the English Revolution, ed. Michael Braddick (Oxford: Oxford University Press, 2015)

Parker, Geoffrey, 'Crisis and Catastrophe: The Global Crisis of the Seventeenth Century Reconsidered', *American Historical Review* 113, no. 4 (2008)

Parker, Geoffrey, *Global Crisis: War, Climate Change and Catastrophe in the Seventeenth Century* (New Haven: Yale University Press, 2013)

Pastore, Christopher L., *Between Land and Sea: The Atlantic Coast and the Transformation of New England* (Cambridge, MA: Harvard University Press, 2014)

Patterson, K. David, 'Yellow Fever Epidemics and Mortality in the United States, 1693–1905', *Social Science and Medicine* 34 (1992)

Peña, Carmen, and Fernando Girón, *La prevención de la enfermedad en la España bajo medieval* (Granada: Universidad de Granada, 2006)

Petegorsky, D.W., *Left-Wing Democracy in the English Civil War* (London: Gollancz, 1940)

Pfister, Christian, 'Climatic Extremes, Recurrent Crises and Witch Hunts: Strategies of European Societies in Coping with Exogenous Shocks in the Late Sixteenth and Early Seventeenth Centuries', *Medieval History Journal* 10 (2006)

Pfister, Christian, 'Weather, Climate, and the Environment', in *Oxford Handbook of Early Modern European History, 1350–1750, vol. 1: Peoples and Place*, ed. Hamish Scott (Oxford: Oxford University Press, 2015)

Pincus, Steven, 'John Evelyn: Revolutionary', in *John Evelyn in His Milieu*, ed. John Hunter and Francis Harris (London: The British Library, 2003)

Pinna, Mario, *La teoria dei climi. Una falsa dottrina che non muta da ippocrate a Hegel* (Rome: Società geografica italiana, 1988)

Pinna, Mario, 'Un Aperçu historique de "la théorie des climats"', *Annales de Géographie* 98, no. 547 (1998)

Pomeranz, Kenneth, 'War, Weather, and Welfare: Persistence and Change in Geoffrey Parker's Global Crisis', *Historically Speaking* 14, no. 5 (2013)

Pooley, Simon, *Burning Table Mountain: An Environmental History of Fire on the Cape Peninsula* (Basingstoke: Palgrave Macmillan, 2014)

Power, Eileen E., 'Pierre Dubois and the Domination of France', in *The Social and Political Ideas of Some Great Medieval Thinkers*, ed. F.J.C. Hearnshaw (London: Harrap & Company, 1923)

Prince, Hugh, *Wetlands and the American Midwest: A Historical Geography of Changing Attitudes* (Chicago: Chicago University Press, 1997)

Pritchard, Sara, *Confluence: The Nature of Technology and the Remaking of the Rhone* (London: Harvard University Press, 2011)

Quenet, Grégory, *Versailles, une histoire naturelle* (Paris: La Découverte, 2015)

Race and Blood in the Iberian World, ed. Max S. Hering Torres, María Elena Martínez, and David Nirenberg (Vienna: LIT Verlag, 2012)

Rawcliffe, Carole, *Urban Bodies: Communal Health in Late Medieval English Towns and Cities* (Woodbridge: The Boydell Press, 2013)

Reappraisals of the Scientific Revolution, ed. David C. Lindberg and Robert S. Westman (Cambridge: Cambridge University Press, 1990)

Reese, Ronald, 'Under the Weather: Climate and Disease, 1700–1900', *History Today* 46 (1996)
Rereading the Black Legend: The Discourses of Religious and Racial Difference in the Renaissances Empires, ed. Margaret Greer, Walter Mignolo, and Maureen Quilligan (Chicago: University of Chicago Press, 2007)
Reuss, Martin, *Designing the Bayous: The Control of Water in the Atchafalaya Basin, 1800–1995* (College Station, TX: Texas A&M University Press, 2004)
Reynard, Pierre, 'Public Order and Privilege: Eighteenth-Century French Roots of Environmental Regulation', *Technology and Culture* 43 (2002)
Richards, John F., *The Unending Frontier: An Environmental History of the Early Modern World* (Berkeley: University of California Press, 2003)
Rippon, Stephen, '"Uncommonly Rich and Fertile" or "Not Very Salubrious"? The Perception and Value of Wetland Landscapes', *Landscapes* 1 (2009)
Roger, Philippe, *The American Enemy: A Story of French Anti-Americanism*, trans. Sharon Bowman (Chicago: University of Chicago Press, 2005)
Root, Hilton, *Peasants and King in Burgundy: Agrarian Foundations of French Absolutism* (Berkeley: University of California Press, 1987)
Rose, Paul L., *Bodin and the Great God of Nature: The Moral and Religious Universe of a Judaiser* (Geneva: Droz, 1980)
Rose, Paul L., 'Bodin's Universe and Its Paradoxes: Some Problems in the Intellectual Biography of Jean Bodin', in *Politics and Society in Reformation Europe: Essays for Sir Geoffrey Elton on His Sixty-Fifth Birthday*, ed. E.I. Kouri and Tom Scott (New York: Macmillan, 1987)
Rosenthal, Jean-Laurent, *Fruits of Revolution: Property Rights, Litigation and French Agriculture 1700–1860* (Cambridge: Cambridge University Press, 1992)
Rowland, C., 'Gerrard Winstanley: Man for All Seasons', *Prose Studies* 36, no. 1 (2014)
Rozwadowski, Helen, and Michael S. Reidy, 'Spaces in between: Science, Ocean, Empire', *Isis* 105, no. 2 (2014)
Rudwick, Martin J.S., *Bursting the Limits of Time: The Reconstruction of Geohistory in the Age of Revolution* (Chicago: University of Chicago Press, 2005)
Rudwick, Martin J.S., 'The Early Geological Society in Its International Context', *The Geological Society* (Special Issue: The Making of the Geological Society of London) 317 (2009)
Rudwick, Martin J.S., *Earth's Deep History: How It Was Discovered and Why It Matters* (Chicago: University of Chicago Press, 2014)
Russell, Conrad, *The Causes of the English Civil War* (Oxford: Clarendon Press, 1990)
Russell, Conrad, *Unrevolutionary England: 1603–42* (London: Hambledon, 1990)
Russell, D., 'Locke on Land and Labour', *Philosophical Studies* 117, nos. 1–2 (2004)
Sabatier, Gérard, *Versailles ou la figure du roi* (Paris: Fayard, 1999)
Sabin, Paul, 'Rooting around in Search of Causality', *Environmental History* 10, no. 1 (2005)
Schiffman, Zachary, *The Birth of the Past* (Baltimore: Johns Hopkins University Press, 2012)
Schmid, Martin, 'The Environmental History of Rivers in the Early Modern Period', in *An Environmental History of the Early Modern Period: Experiments and Perspectives*, ed. Martin Knoll and Reinhold Reith (Berlin: Lit, 2014)
Schneer, Jonathan, *The Thames: England's River* (London: Abacus, 2005)
Schwartz, Stuart B., *Sea of Storms: A History of Hurricanes in the Greater Caribbean from Columbus to Katrina* (Princeton: Princeton University Press, 2015)
Science and Empire in the Atlantic World, ed. Nicholas Dew and James Delbourgo (New York: Routledge, 2008)
Senellart, Michel, *Machiavélisme et raison d'état XIIIe-XVIIIe siècle: suivi d'un choix de textes* (Paris: Presses Universitaires de France, 1989)

Sharp, B., *In Contempt of All Authority: Rural Artisans and Riot in the West of England, 1586–1660* (Berkeley: University of California Press, 1980)

Sharp, B., 'Rural Discontents and the English Revolution', in *Town and Countryside in the English Revolution*, ed. H.C. Richardson (Manchester: Manchester University Press, 1992)

Silverblatt, Irene, 'Foreword', in *Imperial Subjects: Race and Identity in Colonial Latin America*, ed. Andrew Fisher and Matthew O'Hara (Durham: Duke University Press, 2009)

Sir Joseph Banks: A Global Perspective, ed. R.E.R. Banks (Richmond: Royal Botanic Gardens, 1994)

Siraisi, Nancy, *Medieval and Early Renaissance Medicine: An Introduction to Knowledge and Practice* (Chicago: University of Chicago Press, 1990)

Skelton, Leona, *Sanitation in Urban Britain, 1560–1700* (London: Routledge, 2016)

Slack, Paul, 'Books of Orders: The Making of English Social Policy, 1577–1631', *Transaction of the Royal Historical Society* 5th series 30 (1980)

Slack, Paul, *The Invention of Improvement: Information and Material Progress in Seventeenth-Century England* (Oxford: Oxford University Press, 2014)

Slavin, Philip, 'The Great Bovine Pestilence and Its Economic and Environmental Consequences in England and Wales, 1318–50', *Economic History Review* 65, no. 4 (2012)

Smout, T.C., *Nature Contested: Environmental History in Scotland and Northern England since 1600* (Edinburgh: Edinburgh University Press, 2000)

Snider, Alvin, 'Hard Frost, 1684', *Journal for Early Modern Cultural Studies* 8, no. 2 (2008)

Sörlin, Sverker, 'Reconfiguring Environmental Expertise', *Environmental Science & Policy* 28 (2013)

Sörlin, Sverker, and Paul Warde, 'Making the Environment Historical: An Introduction', in *Nature's End: History and the Environment*, ed. Sverker Sörlin and Paul Warde (Basingstoke: Palgrave Macmillan, 2009)

Spada, F., and S. Passigli, 'Paesaggio, risorse e percezione dell'ambiente nelle zone umide: esempi dall'area romana medievale (secoli X–XV)', in *Les zones humides méditerranéennes hier et aujourd'hui / Le zone umide mediterranee ieri e oggi*, ed. Magalie Franchomme, Christine Labeur, Daria Quatrida, and Remy Simonetti (Padua: Padova University Press, 2014)

Spowers, R., *Rising Tides* (Edinburgh: Canongate, 2002)

Stamp, Tom, and Cordelia Stamp, *William Scoresby, Arctic Scientist* (Whitby: Caedmon of Whitby Press, 1976)

Staszak, Jean-François, *La géographie d'avant la géographie. Le climat chez Aristote et Hippocrate* (Paris: L'Harmattan, 1995)

Stewart, Laura A.M., *Rethinking the Scottish Revolution: Covenanted Scotland 1637–51* (Oxford: Oxford University Press, 2016)

Stine, Jeffrey K., *America's Forested Wetlands: From Wasteland to Valued Resource* (Durham, NC: Forest History Society, 2008)

Suire, Yannis, *Le marais Poitevin. Une écohistoire du XVIe siècle à l'aube du XXe siècle* (La Roche sur Yon: Centre vendéen de recherches historiques, 2006)

Sutton, Philip W., *The Environment: A Sociological Introduction* (Cambridge: Polity Press, 2007)

Sweet, James, 'The Iberian Roots of American Racist Thought', *William and Mary Quarterly* 54, no. 1 (1997)

Swope, Kenneth M., *The Military Collapse of China's Ming Dynasty, 1618–44* (New York: Routledge, 2014)

Tapsell, Grant, *The Person Rule of Charles II, 1681–85* (Woodbridge: Boydell, 2007)

Taylan, Ferhat, 'La Rationalité mésologique. Connaissance et gouvernement des milieux de vie (1750–1900)' (PhD thesis, Université de Bordeaux III, Bordeaux, 2014)

Thomas, Keith, *Man and the Natural World: Changing Attitudes in England, 1500–1800* (London: Penguin, 1983)

Thompson, E.P., 'The Moral Economy of the English Crowd in the Eighteenth Century', *Past & Present* 50 (1971)

Thompson, E.P., *Whigs and Hunters: The Origin of the Black Act* (London: Penguin Books, 1975)

Thornton, Russell, *American Indian Holocaust and Survival: A Population History since 1492* (Norman: University of Oklahoma Press, 1987)

Tooley, Marian J., 'Bodin and the Mediaeval Theory of Climate', *Speculum* 28, no. 1 (1953)

Tudela de la Orden, José, 'Economía', in *El legado de España a América*, ed. José Tudela de la Orden (2 vols; Madrid: Ediciones Pegaso, 1954)

Urban, Michael A., 'An Uninhabited Waste: Transforming the Grand Prairie in Nineteenth Century Illinois, USA', *Journal of Historical Geography* 31 (2005)

Valenčius, Conevery Bolton, *The Health of the Country: How American Settlers Understood Themselves and Their Land* (New York: Basic Books, 2002)

Van Dam, Petra, and Milja van Tielhof, *Waterstaat in Stedenland. Het Hoogheemraaschap van Rijnland voor 1857* (Utrecht: Matrijs, 2007)

Van der Eijk, Philip J., 'Airs Waters Places and on Sacred Disease: Two Different Religiosities?', *Hermes* 119, no. 2 (1991)

Vileisis, Ann, *Discovering the Unknown Landscape: A History of America's Wetlands* (Washington: Island Press, 1997)

Viñas Mey, Carmelo, 'Datos para la historia económica de la colonización española', *Revista nacional de economía* 44 (1923)

Vogel, Brant, 'The Letter from Dublin: Climate Change, Colonialism, and the Royal Society in the Seventeenth Century', *Osiris* 26 (2011)

Walsham, Alexandra, *Providence in Early Modern England* (Oxford: Oxford University Press, 1999)

Walter, John, *Crowds and Popular Politics in Early Modern England* (Manchester: University of Manchester Press, 2010)

Warde, Paul, *Ecology, Economy and State Formation in Early Modern Germany* (Cambridge: Cambridge University Press, 2006)

Warde, Paul, 'Fear of Wood Shortage and the Reality of the Woodland in Europe, c. 1450–1850', *History Workshop Journal* 62 (2006)

Warde, Paul, 'The Environmental History of Pre-Industrial Agriculture in Europe', in *Nature's End: History and the Environment*, ed. Sverker Sörlin and Paul Warde (Basingstoke: Palgrave Macmillan, 2009)

Warde, Paul, 'The Idea of Improvement, c.1520–1700', in *Custom, Improvement and the Landscape in Early Modern Britain*, ed. R. Hoyle (Farnham: Ashgate, 2011)

Warde, Paul, 'The Invention of Sustainability', *Modern Intellectual History* 8, no. 1 (2011)

Warde, Paul, 'Imposition, Emulation and Adaptation: Regulatory Regimes in the Commons of Early Modern Germany', *Environment and History* 19, no. 3 (2013)

Warde, Paul, 'Crisis or Coincidence?', *Past & Present* 228 (2015)

Watson, Fiona, 'Rights and Responsibilities: Wood Management as Seen through Baron Court Records', in *Scottish Woodland History*, ed. T.C. Smout (Edinburgh: Edinburgh University Press, 1997)

Wear, Andrew, *Knowledge and Practice in English Medicine, 1550–1680* (Cambridge: Cambridge University Press, 2000)

Wear, Andrew, 'Place, Health, and Disease: The Airs, Waters, Places Tradition in Early Modern England and North America', *Journal of Medieval and Early Modern Studies* 38, no. 3 (2008)

Webb, D., 'The Bitter Product of Defeat? Reflections on Winstanley's Law of Freedom', *Political Studies* 52 (2004)

Wheeler, Roxanne, *The Complexion of Race: Categories of Difference in Eighteenth-Century British Culture* (Philadelphia: University of Pennsylvania Press, 2000)

White, Richard, '"Are You an Environmentalist or Do You Work for a Living?": Work and Nature', in *Uncommon Ground: Toward Reinventing Nature*, ed. William Cronon (New York: W.W. Norton & Company, 1995)

White, Richard, *The Organic Machine: The Remaking of the Columbia River* (New York: Hill & Wang, 1995)

White, Sam, *The Climate of Rebellion in the Early Modern Ottoman Empire* (Cambridge: Cambridge University Press, 2013)

White, Sam, 'The Real Little Ice Age', *Journal of Interdisciplinary History* 44, no. 3 (2014)

White, Sam, 'Unpuzzling American Climate: New World Experience and the Foundations of a New Science', *Isis* 106 (2015)

Wickham, Gary, and Jo-Ann Goodie, *Legal and Political Challenges of Governing the Environment and Climate Change: Ruling Nature* (London and New York: Routledge, 2013)

Wickman, Thomas M., 'Snowshoe Country: Indians, Colonists, and Winter Spaces of Power in the Northeast, 1620–1727' (PhD dissertation, Harvard University, Cambridge, MA, 2012)

Williams, Glyndwr, *Voyages of Delusion: The Search for the Northwest Passage in the Age of Reason* (New Haven: Yale University Press, 2003)

Williams, Michael, 'Agricultural Impacts in Temperate Wetlands', in *Wetlands: A Threatened Landscape*, ed. Michael Williams (Oxford: Basil Blackwell, 1990)

Williams, R., *Problems in Materialism and Culture* (London: Verso, 1980)

Wilson, Anthony, *Shadow and Shelter: The Swamp in Southern Culture* (Jackson: University Press of Mississippi, 2005)

Winchester, Angus, *The Harvest of the Hills: Rural Life in Northern England and the Scottish Borders, 1400–1700* (Edinburgh: Edinburgh University Press, 2000)

Winiwarter, Verena, 'The Art of Making the Earth Fruitful: Medieval and Early Modern Improvements in Soil Fertility', in *Ecologies and Economies in Medieval and Early Modern Europe*, ed. Scott G. Bruce (Leiden: Brill, 2010)

Wood, Andy, *Riot, Rebellion, and Popular Politics in Early Modern England* (New York: Palgrave Macmillan, 2002)

Wood, Andy, *The 1549 Rebellions and the Making of Early Modern England* (Cambridge: Cambridge University Press, 2007)

Wood, Gillen D'Arcy, *Tambora: The Eruption That Changed the World* (Princeton: Princeton University Press, 2014)

Wood, Gordon, *The Rising Glory of America, 1760–1820* (rev. edn; Boston: Northeastern University Press, 1990)

Wood, Gordon, 'Environmental Hazards, Eighteenth-Century Style', in *Old World, New World: American and Europe in the Age of Jefferson*, ed. Leonard J. Sadosky, Peter Nicolaisen, Peter S. Onuf, and Andrew O'Shaughnessy (Charlottesville: University of Virginia Press, 2010)

Worster, Donald, 'Doing Environmental History', in *The Ends of the Earth: Perspectives on Modern Environmental History*, ed. Donald Worster (Cambridge: Cambridge University Press, 1988)

Worster, Donald, *Nature's Economy: A History of Ecological Ideas* (Cambridge: Cambridge University Press, 1994)

Wright, Peter, *Life on the Tyne: Water Trades on the Lower River Tyne in the Seventeenth and Eighteenth Centuries, a Reappraisal* (Farnham: Ashgate, 2014)

Wrightson, Keith, *English Society 1580–1680* (New Brunswick: Rutgers University Press, 1982)

Wrightson, Keith, *Ralph Tailor's Summer: A Scrivener, His City and the Plague* (London: Yale University Press, 2011)

The Year Without a Summer? World Climate in 1816, ed. C.R. Harrington (Ottawa: Canadian Museum of Nature, 1992)

Yonge, C.D., *The History of the British Navy from the Earliest Period to the Present Time* (2 vols; London: Richard Bentley, 1863)

Zalasiewicz, Jan, and Mark Williams, *Ocean Worlds: The Story of Seas on Earth and Other Planets* (Oxford: Oxford University Press, 2014)

Zilberstein, Anya, 'Inured to Empire: Wild Rice and Climate Change', *William and Mary Quarterly* 72, no. 1 (2015)

Zilberstein, Anya, *A Temperate Empire: Making Climate Change in Early America* (Oxford: Oxford University Press, 2016)

Zuidervaart, Huib J., 'An Eighteenth-Century Medical-Meteorological Society in the Netherlands, Instrumentation and Quantification: Part 2', *British Journal for the History of Science* 39 (March 2006)

Zupko, Ronald, and Robert Laures, *Straws in the Wind: Medieval Urban Environmental Law: The Case of Northern Italy* (Oxford: Westview, 1996)

Web links

Gopnik, Adam, 'Faces, Places, Spaces: The Renaissance of Geographical History', *The New Yorker*, www.newyorker.com/magazine/2012/10/29 Last accessed 10 February 2016

Monbiot, G., 'Still Digging', June 2000 www.monbiot.com/archives/2000/06/01/still-digging/ Last accessed 13 June 2016

Rahm Emmanuel Speech to the Wall Street Journal CEO Council, November 20, 2008, podcast.mktw.net/wsj/audio/20081120/pod-wsjemmanuel/pod-wsjemmanuel.mp3 Last accessed 26 February 2016

Index

Note: End note information is denoted by an n and note number following the page number.

For Product Safety Concerns and Information please contact our EU representative GPSR@taylorandfrancis.com Taylor & Francis Verlag GmbH, Kaufingerstraße 24, 80331 München, Germany